AIRFIX
magazine guide 15

World War 2 Wargaming

Bruce Quarrie

Patrick Stephens Ltd
in association with Airfix Products Ltd

First published — 1976

ISBN 0 85059 230 5

Cover design by Tim McPhee

Text set in 8 on 9 pt Helvetica Medium by Blackfriars Press Limited, Leicester.
Printed on Fineblade cartridge 90 gsm and bound by the Garden City Press, Letchworth, Herts.
Published by Patrick Stephens Limited, Bar Hill, Cambridge, CB3 8EL, in association with Airfix Products Limited, London SW18.

Contents

Acknowledgements

Once again, several people have contributed to formulating this book, not least of whom is my wife Linda and some of her pupils from the Manor School, Cambridge, who made many of the models illustrated in the photographs: Steven Alsop, David Barber, Philip Cater, Barrie Crick, Douglas Dack, Kevin Diss, Gary Hall, John Hayden, Robert Lee, Alison and Paul Loker, Colin Long, Deborah Patman, John Rowley, Nicola Stokes and Robert Thompson. Special thanks also to Jeff Parker for providing much of the scenery and taking many of the photographs, Verity Galloway and Chris Mills for providing transport, and Gerald Scarborough for giving me permission to use a number of his photographs.

Editor's introduction

Modern wargaming, using model tanks, guns and soldiers, has appealed to millions of youngsters — myself included — from the day they got their first Dinky Centurion tank for Christmas. Our garden, when I was nine or ten, was an armed fortress, my brother having one end for his airfields, missile sites, tank depots, barracks and underground HQs, I the other. Because I was older I got the end with the stream, so I could have a harbour too!

Those were the glorious, unsophisticated days. It didn't matter to us that we were using heat-seeking missiles to shoot down Hurricanes or Lancasters, or that we quite happily pitted 1:32 scale plastic tanks against far smaller models. We didn't know much about weapons or firepower, so casualties were usually done on the basis of, 'well, mine's bigger than yours, so I win.' Probably most readers of this book have similar experiences.

That sort of wargaming is all very well, and great fun (especially when you get older, acquire an air gun, and can proceed to knock merry hell out of your opponent's models!) but it can hardly be called realistic.

In this book, I have attempted to show how correctly scaled and organised forces from all the major combatant nations of World War 2 can be constructed and made to fight against each other with the help of a tape measure and dice. Due to lack of space, aerial combat has had to be excluded, but this book provides all the essential information you need to begin moving and fighting with model soldiers, tanks and guns of various types, either on a proper wargames table, on the floor, or even in the garden or on the beach. All you need are a few Airfix models and figures, and you can begin.

Infantry weapons, tanks and self-propelled guns, armoured cars and artillery pieces are all covered, and you can add to and adapt the playing rules as you go along to create even greater realism if you wish.

To the experienced wargamer who already specialises in this period I would just say one thing. This book is intended to provide a fast-moving and exciting game without bogging the players down in too many technicalities. The playing rules are the result of several years' trial and error experimentation, and work well for me. So even if you normally prefer a more sophisticated game, give them a try.

To the complete beginner I would say this: wargamers invariably disagree about practically everything, and especially sets of rules. So if you join a local wargames club, you will probably find as many sets of rules in operation as there are members. But if you've got the hang of these first, you'll easily be able to adapt to different ones, because the basic principles are essentially all the same.

Good battling!

BRUCE QUARRIE

one

Strictly for beginners

Mention the word 'wargaming' to most people and they will usually say something vague like, 'Oh yes. Playing with toy soldiers. Don't you use dice, or something?' If you are already a wargamer, you're undoubtedly familiar (and fed up!) with this. But this chapter is aimed at those readers who are still a bit vague as to how wargaming actually works. If this is all old hat to you, turn to chapter two.

On the simplest level, wargaming begins when children start firing matchsticks at their Woolworth's cowboys, indians, knights in armour, Guardsmen and other assorted figures. This can be quite fun if tackled a little more systematically, and indeed the novelist H. G. Wells once wrote a complete book on how to do it. Called *Little Wars,* this famous book has recently been re-issued by an enterprising publisher, and is well worth borrowing from your local library.

Once you get tired of firing matchsticks, however — which soon happens when you see what happens to your soldiers' paintwork — you begin wondering how you can simulate combat without actually firing physical objects at your troops.

At the very simplest level, you could throw a dice for each soldier firing, and allow that every six scored kills one of the opposing men. But this relegates wargaming to a pure game of chance, and does not take into account weapon ranges, rates of fire, type of ammunition, or whether the target is moving, lying down or sheltering behind a stone wall, etc.

Wargames rules are usually devised to take these factors into account. They operate on a probability basis — that is, if you know from historical records that a particular gun was 66 per cent accurate at a given range, you would write a rule saying that it would score a hit every time you threw a 3, 4, 5 or 6 on the dice, giving you four chances out of six, or 66 per cent. But since you also know that a moving target is more difficult to hit than a stationary one, you would also have a rule saying that if the target was moving, you would deduct one from your dice score — so a 3 would be a miss. And so you go on.

Movement in a wargame is a question of scaling down known troop march rates and vehicle speeds to something feasible in a limited space, in the same way that a map is a scaled down representation of a stretch of countryside. Thus, for example, you

A troop of 25 pdrs in action against PzKpfw IVs, Western Desert, 1941.

could have an inch representing ten yards, and work out your permissible moves from that.

You also need a time scale, for pretty obvious reasons. It's all very well saying that a man can march at three miles per hour, for example, and that this means he can march 528 inches per hour (if one inch is being made to equal ten yards). But 528 inches is nearly 15 yards — far more than the length of the average wargames table. So we have to have a Move time scale which will allow reasonable movement within the restrictions of the playing area. This time scale also determines how many times weapons can be fired during a Move, etc.

Once you have established basic ground and time scales, everything else falls in place. You can work out how far and how fast men and vehicles can travel during a wargames 'Move'; how many times they can fire their guns, and at what ranges; and, using historical records and weapon data tables from specialist books, what the probabilities of hits are under different circumstances.

That's what this book is about. It's not the last word by any means, because one of the big problems — and also one of the major attractions — of World War 2 as a wargaming period is the vast number of different vehicles and weapons employed by the combatant nations, and it is impossible to take every single factor into account in a book this size. There *are* some sets of rules available which do go into tremendous detail — and practically need a computer to work out! Although these types of rules give highly realistic games, in which practically every relevant factor is considered, they also produce very slow-moving ones, in which it could take ten minutes to calculate the effect of one gun firing at one target.

To my mind this is not war *gaming.* My dictionary defines 'game' as a 'contest for amusement', and I have always regarded the enjoyment of a good game as far more important than one hundred per cent realism. So in this book I have attempted to give some playing ideas and rules which will enable you to fight fast-moving and exciting battles, with reasonably realistic results.

As you become involved in the subject, you will probably want to find out more about the fighting forces involved and the weapons and vehicles used during World War 2, so I've included a few suggestions for further reading. Sooner or later you will almost certainly come to disagree with some of my interpretations, and start re-writing the rules for yourself. Once you do that, you are a true wargamer: we're a pig-headed breed, and nobody else's rules are ever right! But in the meanwhile, I trust that this book will serve as a useful introduction to World War 2 wargaming for anybody who hasn't tackled it before, while at the same time I hope that some of the 'old hands' will find some of my ideas useful. Give them a chance anyway.

two

Wargame models

World War 2 is undoubtedly one of the most popular periods with wargamers, largely thanks to the vast range of vehicles and figures in a multiplicity of scales which is available. However, this very variety makes it difficult for the would-be wargamer to know where to begin, and these introductory remarks are aimed at those readers who would like to know some of the choices open to them before committing themselves.

The first problem is scale. Model vehicles and figures suitable for wargaming can be purchased in anything from 1:300 to 1:32 scale. The former are tiny models, frequently half an inch or less in length, cast in metal with varying degrees of detail and accuracy. In many ways these are the ideal models for World War 2 wargaming because their small size means you can deploy large numbers on the average wargames table, thereby more accurately representing the large and varied numbers of troops and vehicles present during a battle in this period. The 1:32 scale models (including 1:35) are large plastic construction kits and polystyrene, polythene or metal figures which go with them. These are more suited for display on their own or inclusion in dioramas than for wargaming, although interesting small-scale encounters between, say, two infantry sections can be fought using them.

Readers who are attracted by either of the above should read *Tank Battles in Miniature,* by Donald Featherstone, which deals with the Western Desert Campaign using 1:300 scale micro-armour; *Tank Battles in Miniature 2,* by myself, which deals with the Russian Campaign in the same scale; or Donald Featherstone's more recent book, *Skirmish Wargaming,* which deals with small-scale battles using just a few figures on a man-to-man basis. A third 'Tank Battles' book, dealing with the D-Day and Normandy campaigns of 1944, is under preparation at the time of writing.

In between 1:300 and 1:32 lies a vast selection of alternative scales — 1:200, 1:150, 1:87, 1:48 and, of course, the most popular of all, 1:76/1:72 scale. 1:76 corresponds to the standard model railway gauge of 4 mm to 1 foot, while 1:72 is model aircraft 'standard scale' or 1 inch to 6 feet. The two scales are very close (1 mm difference

Modifications to existing kits can be simple and add variety to your games. The Russian T-34/85 bulldozer was modelled by Chris Ellis.

in every scale 6 feet of model), and although pure modellers shudder at the thought of mixing them in a constant scale collection, most wargamers combine them quite happily — and, indeed, also include the 1:87 scale Minitank models in order to get miniature representations of vehicles not otherwise available.

The popularity of 1:76 for wargaming began many years ago when Airfix first began releasing their sets of 00/HO figures and model tank and gun kits, and has increased in proportion to the growing availability of different figures and models, latterly from firms such as Eidai, ESCI, Fujimi, Hasegawa, Matchbox and Nitto as well as Airfix. Because of its widespread acceptance, I am assuming in this book that you will also want to use 1:76 scale models, and the playing rules are geared to this. However, they are easily adaptable to other scales if you wish.

It would be impossible to list all the vehicles and figures currently available for wargamers in 1:76 scale, since this would run to about two pages and in any case, new kits are being released so frequently that such a list would be out of date by next week. It is sufficient to say that practically every major tank used during the last war is readily available in kit form, together with a growing selection of armoured cars, self-propelled guns, artillery pieces, half-tracks and 'soft-skin' trucks. In most cases gaps in the range can quite easily be filled by converting existing models into different vehicles, or by scratch-building using polystyrene sheet (plastic card), scrap 'sprue' and odd left-over parts from kits. Large numbers of such projects have been and still are regularly covered in *Airfix Magazine, Military Modelling* and the MAFVA magazine *Tankette.* (The MAFVA — Miniature Armoured Fighting Vehicle Association — is an invaluable society for any World War 2 wargamer to join, because not only does membership entitle you to the magazine, but it also enables you to purchase the special glass fibre model kits in 1:76 scale which are privately made for members only. For further details write, enclosing a stamped, self-addressed envelope, to: Gary Williams, 15 Berwick Avenue, Heaton Mersey, Stockport, Cheshire SK4 3AA.)

Apart from the above-mentioned magazines, a large number of conversion and scratch-building projects are covered in Gerald Scarborough's two books in this series, number 3 — *Military Modelling,* and number 5 — *Tank & AFV Modelling,* while scale plans for dozens more can be found in the little Bellona booklets.

A good example of a scratch-built addition to a wargames army is this Chevrolet 4×4 truck made by Canadian Airfix Magazine *reader Harold Morrison.*

This Daimler armoured car is a typical example of the quality of the glass fibre kits which can be purchased by members of the Miniature Armoured Fighting Vehicle Association.

Early Tiger 1 constructed from a Hasegawa 1:72 scale kit (Gerald Scarborough).

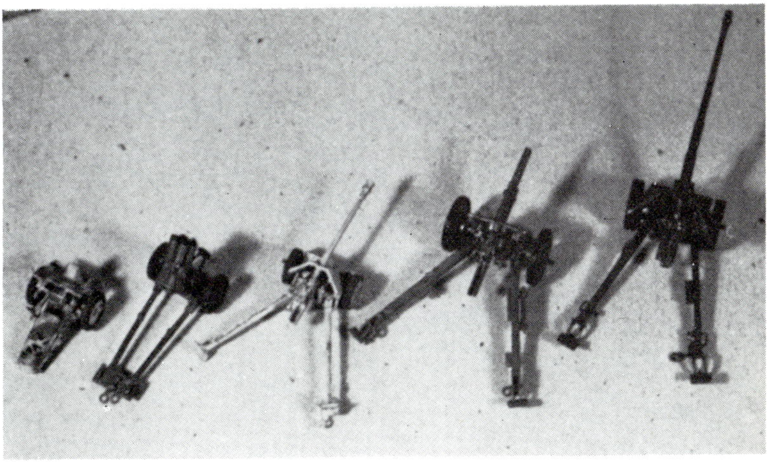

The cast metal guns manufactured by Hinchliffe are valuable additions to any wargamer's arsenal. Pictured are (l-r) a German 7.5 cm infantry howitzer, Nebelwerfer and 7.5 cm Pak 40, British 17 pdr and German 8.8 cm Pak 41.

Model construction

Because tank and other vehicle models for wargames use invariably come in for a great deal more handling than a kit which is built up to go into a pure model collection, a slightly different modelling technique is required.

The first essential in a wargames model is robustness, and while the majority of kits will stand a fair amount of handling if constructed properly, there are several simple additional precautions you can take to make them even stronger. To begin with, flimsy external detail parts such as headlights and radio aerials can be left off entirely. So long as the finished model clearly represents a particular vehicle, I do not feel that details of this nature are important in a wargame. This does not mean that the external surfaces need to be left entirely bare. The parts of a model tank which are going to take the most wear are the hull and turret sides, so small detail

'The two best tanks of the war', a claim which is often made for the Russian T-34 (above) and the German Panther (below) whose design was heavily influenced by the former.

(smoke dischargers, tools, tow chains, etc) should be left off. However, the hull front and rear will hardly ever be touched, so there is no reason why spare track links, rolls of camouflage netting and similar accessories should not be cemented firmly in these areas to give your vehicles a slightly more individual appearance.

The weakest part of most small tank models is the wheels and tracks. If you are not too bothered by the prospects of road wheels or drive sprockets coming loose (and they are the devil's job to glue back in position if the axle stub has broken off in its locating hole) then simply cement all these in position very firmly using an epoxy resin rather than polystyrene cement. Polystyrene cement achieves its bond by actually melting, or welding, the component surfaces together, but glue on the flimsy axle stubs often softens them, resulting in their bending or breaking off entirely at the slightest excuse. Epoxy resin does not have this effect.

The preferable alternative is to drill out the centres of all the wheels and fix them on to the model by means of stiff piano wire axles cemented right through the hull. It is not always possible to do this (for example, on the Sherman), but some protection can usually be given to the trackwork by positioning track guards made from fairly thick plastic card on the vehicle's sides.

Internal strengthening can easily be achieved on most models by inserting shaped pieces of balsa wood inside the vehicle during assembly, while many wargamers also add strips of lead or balls of Plasticine inside their models to give them a bit more weight. Tanks are heavy beasts and it seems slightly incongruous to have a Tiger tank which could be blown away in a strong breeze! This is entirely up to you, however.

Crew figures always cause a certain amount of debate. Some wargamers prefer to build their models battened down with nobody showing, and thereby run the risk of some crafty opponent claiming that their visibility must be restricted; while others like having crew figures all over the place and run the risk of having them shot by snipers!

Historically speaking, the only crew member who would normally be visible in a tank going into battle was the commander, who 'conned' the vehicle from the turret hatch or cupola, issuing directions to the driver (whose vision was very limited) and 'spotting' for the gunner. In practice, too, it is almost impossible to fight a tank while completely closed down, and in most armies during World War 2 the commander would fight with his head at least poking out, and this is how I model most of my own vehicles. For their own safety tank commanders were normally told to wear steel helmets, but in practice they frequently did not because they were extremely uncomfortable to wear over a pair of headphones. Once again, the choice is up to you.

Finally, some wargamers mount their model vehicles on cardboard bases to minimise handling, and there is no reason why you should not do this, except that it always looks rather unrealistic to see a tank on a green base haring along a grey or brown road!

Figures

Wargames figures come in for even more handling than the vehicles, which is unfortunate since the soft polythene of which the Airfix figures are made does not 'take' paint particularly well. To minimise the effects of constant handling, a number of precautionary measures can be taken.

First, wash all your figures thoroughly in a solution of washing-up liquid, then dry them carefully. Keep

Matchbox Sherman Firefly on its diorama base. Buy two kits and you have two broken bridge sections which are ideal for use on a wargames table.

your bare fingers off the figure itself — hold it by the base or in a pair of tweezers, since no matter how clean your hands are they are always slightly greasy, and this grease rapidly gets transferred back to the figure.

Now undercoat the figure. Some modellers swear by an undercoat of a white glue such as Unibond, but I prefer to use ordinary gloss varnish. Either treatment has the twofold effect of providing a surface on which paint will adhere properly, and lending the figure additional rigidity.

Once the primer is dry a second undercoat of matt white paint should be applied. This is not strictly necessary, but I usually do it because it helps bring out colours better.

Painting in the appropriate uniform colours comes next, after which the figure should be given a second coat of varnish, this time using Airfix MV1 matt varnish. This needs very thorough stirring as it tends to go glossy in patches, and should preferably be applied in two or three thin coats rather than one thick one, to prevent the varnish running and congealing in unsightly greyish patches in all the crevices on the figure.

Finally, all figures should ideally be mounted on new bases of thin strip lead (your local newspaper office will probably be able to let you have some offcuts of the printing metal strips used to 'lead out' type at low cost). This gives them extra stability and prevents them falling over on slopes on the wargames table. Small sub-units such as machine-gun or mortar teams of two or three men can be cemented on to common bases if desired. If you do this though you will need at least two figure groups to represent each machine-gun or mortar crew — one with the figures moving, carrying their weapons, the other in the 'action' position. Many wargamers extend this principle to all their figures, some going so far as to have half a dozen figures for each man (for example, one prone, one kneeling, one standing, one running, one wounded and one dead), but it is not necessary to go to this extreme.

Hinchliffe also make field artillery. Here is a German 155 mm gun.

three

Ground, time and model scales

Having established that we are going to use 1:76 scale vehicles and figures in our wargames, the next essential is to devise some form of basic yardstick for movement and weapon ranges. A moment's thought will clearly show you that one cannot simply use the same ratio of 1:76. A German Tiger 1, for example, had a maximum speed of some 38 kph on a good road. 1/76 of this is 500 metres per hour, so even if our wargames Move time scale was established at one minute per Move, the model Tiger would still be capable of travelling nearly 9 metres per Move — four or five times the length of the average wargames table. Similarly, a German 8.8 cm gun, with an effective anti-tank range of some 1,400 metres, would have a 1:76 scale range of over 18 metres!

It is thus apparent that a ground scale different to the actual model vehicle scale has to be introduced. In many ways this is unfortunate, but given the speed and weapon ranges of World War 2 technology, it is inevitable.

There are a number of different ground scales preferred by various wargamers and rule-writers, of which the most common are either 1 mm to 1 yard (or metre), or 1 inch to 50 yards. The latter is the system advocated in the *Tank Battles in Miniature* books referred to earlier because it is a more appropriate scale for the 1:300 scale vehicles which are the prime concern there. The alternative, 1 mm to 1 metre, is that used by the Wargames Research Group, and who am I to argue with them! The WRG also use a time scale of 30 seconds to a Move which has been designed to allow vehicles a reasonable speed on the table top without their being able to dash from one end of the table to the other in a single bound!

Taking our two previous examples again, the Tiger's maximum road speed of 38 kph translates into roughly 320 metres every 30 seconds, or in other words, 320 mm on the wargames table (that's just over a foot for those readers who still think in inches); while the '88's effective range of 1,400 metres becomes 1,400 mm on the table (just under 5 feet).

30 seconds to a Move is also a convenient time scale when one comes to consider the rapidity of modern firepower. The 2½ minute Move used, for example, in Napoleonic wargaming (see No 4 in this series), is perfectly satisfactory when the average infantryman can only fire about two shots per minute, and his maximum range is only a couple of hundred yards. But when you look at a modern rifle, you find that its rate of fire in trained hands is five or six times this and its effective range three times as great, while automatic weapons shoot even more rapidly.

The next item which needs consideration is the real life to model ratio. Looking back at our Napoleonic example again, where we are concerned with large bodies of infantry and cavalry moving in carefully prescribed formations, it is perfectly acceptable to scale things down and rule that one model soldier represents 33 actual men, so that a battalion of a thousand or so becomes a wargames unit of 20-30 figures. But it would be somewhat incongruous to say that one model Tiger tank represents 33 real vehicles! Thus in modern wargaming it is normal to say that each model vehicle or man represents just one actual vehicle or man, even though this itself creates anomalies in comparison with the ground scale (a model tank 3 inches long by a couple of inches wide, for example, covering a scale ground area of 75 by 50 metres!). However, short of going

American M3 half-tracks advance, France 1944. The gun is a 155 mm howitzer from a Revell helicopter kit.

back to a ground scale 1/76 of the real thing, which we have already settled is impractical, there is no way round this. It doesn't bother most people anyway, because a wargame is still just a highly stylised representation of real war no matter what you do, and any rule writer is forced to compromise between realism and game playability. Try it yourself and see!

To summarise, before moving on to consider World War 2 wargaming in more detail, our basic rules are:

1 mm on the table equals 1 metre in real life
1 Move represents 30 seconds of real-life activity
1 model figure or vehicle represents one real man or vehicle.

four

Choosing and organising an army

A further reason for the popularity of World War 2 with wargamers is the large number of nations involved, each with its own characteristics, uniforms, organisation, weapons and equipment. Again, a detailed study of any one nationality would require far more space than is available here, but there are numerous books in print on the different armies which are readily available from your local public library if you can't afford to buy them.

By far the most popular army is the German. The reasons for this are somewhat obscure, for certainly nobody could claim the Wehrmacht was fighting in a noble cause. Partly it is probably because the Germans fought everybody, so if you've got a German army you can always be sure of finding an opponent — even if you have to match the Wehrmacht against the SS! Partly it is probably fascination with German vehicles — most people have heard of the Tiger or Panther, for example, but comparatively few of the BT-5 or M 13/40. Partly it is probably admiration for the way the small but highly mobile and well organised German army took on the might of Europe and Russia, and very nearly succeeded in 1940-42.

In any case, you will have your own motives for choosing a German army — and I'd be interested to hear them if you care to drop me a line c/o the publishers of this book.

If you do choose a German army, the field is wide open. There are more models of German vehicles and equipment available than for any other nationality, from the tiny PzKpfw I and II, through the Wehrmacht's 'work-horses', the PzKpfw III and IV, up to the famous Tiger 1, Tiger II and Panther, and including most of the self-propelled gun variants, at least three armoured cars, trucks, half-tracks, anti-tank, anti-aircraft and field guns. It is thus possible to build up representative forces for any one or more of the German army's campaigns in Europe, Russia and North Africa. You will need to read other books to discover when the various vehicles entered service, and where they saw combat (for example, Airfix Magazine Guide 8 — *German Tanks of World War 2,* by Terry Gander and Peter Chamberlain), because it would be quite wrong to have King Tigers fighting the French in 1940, to take an extreme example! Later in this chapter I have given an outline to this subject which will act as a guide while you are getting started.

Next most popular is surely the British army, for fairly obvious reasons this time, and once again the wargamer is well served with vehicles, tanks, guns and figures, although at the time of writing there were some noticeable gaps. However, you can buy Crusaders, Matildas, Valentines, Honeys and Lee/Grants for the desert campaign, Shermans, Churchills and Comets for the 1944-5 campaign, a couple of armoured cars, numerous trucks and at least two standard artillery pieces, anti-tank guns (but no anti-aircraft weapons yet), and several sets of figures. A book on British tanks of World War 2, also by Terry Gander and Peter Chamberlain in this series, was under preparation at the time of writing.

Similarly, if you choose an American army, you can buy Lees, Stuarts, Shermans, Chaffees, at least one armoured car, various trucks, an amphibian, half-tracks and a couple of artillery pieces.

When one comes to consider the other major protagonists in World War 2, however, the problems become greater. The Russian army, for example, was the largest of the whole conflict, but there are still relatively few

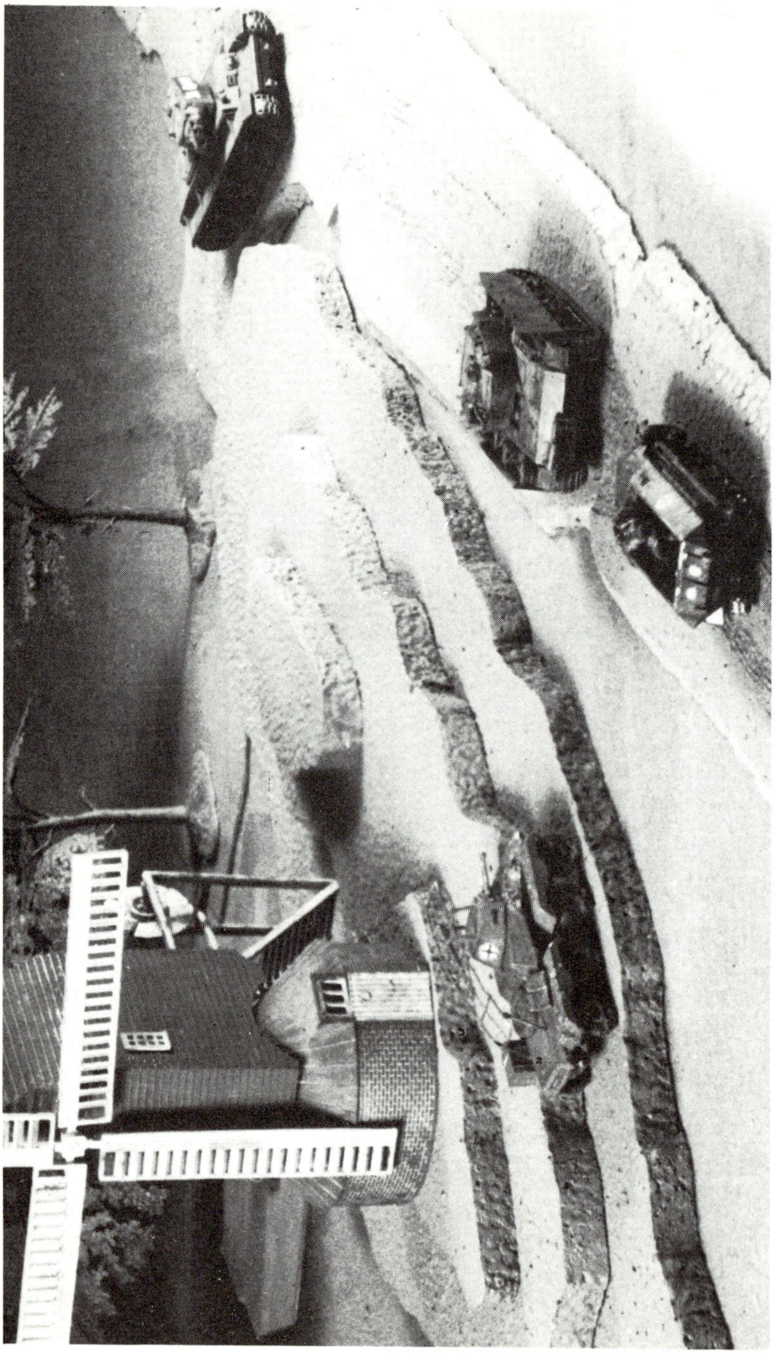

A typical wargames infantry section, in this case German. On the left is the machine-gunner and loader, with sub-machine-gun-armed officer alongside, and seven men with carbines.

models of their tanks apart from the ubiquitous T-34 and KV series. There is one set of figures, one artillery piece and one self-propelled gun, but no armoured cars or trucks at all. An article on modelling the latter appeared in the January 1976 issue of *Airfix Magazine.*

This lack of equipment does not matter so much in the case of the Japanese, since their warfare was principally jungle fighting between infantry, in which tanks were relatively unnecessary. Even so, there are a couple of Japanese tank and truck kits, an infantry gun and two sets of figures. The poor old Italians are hardest done by, since there is not even a set of figures available, although at the time of writing a couple of tank kits were rumoured to be on the way.

Apart from these forces, you could also, if you preferred, choose something like a Free French brigade using British equipment (there are no World War 2 French figures or vehicle kits available at all); or a Polish, Australian, Indian, Canadian, New Zealand or other Allied contingent. *British and Commonwealth Armoured Formations 1919-46,* by Duncan Crow (Profile AFV/Weapons book 2) is a great help in organising these.

Having chosen a nationality, the next problem is how to go about organising it. In practice, each army had its own structure, number of men and type of weapons per unit, rank system and so on, but in effect, until one reaches division or Corps level, they were all very similar.

Infantry

The basic infantry fighting unit was the Section, normally of around ten men commanded by an NCO, equipped principally with rifles or carbines, though with an increasing proportion of automatic weapons as the war progressed.

A typical Section would have the NCO armed with a pistol and/or sub-machine-gun; one light machine-gunner and one loader/ammunition carrier; and seven privates armed with rifles or carbines and hand grenades.

Sections were organised into Platoons, normally three or four Sections to a Platoon, with two to four Platoons forming a Company, and two to four Companies (plus various heavy weapon Sections, Platoons or Companies) forming a Battalion.

For wargames purposes, a Battalion is almost certainly the largest force that will be required on the table. With a hundred men or more to a Company plus the heavy weapons units, etc, an average Battalion is at least 600 and more normally 800 to 1,000 men strong. Since each of these men has to be moved individually on the table (unlike in earlier wargames periods, where troops can be mounted on mul-

Russia, 1943. An SdKfz 232 reconnaissance armoured car shelters behind the crest of a hill from the fire of the KV-1 sitting astride the road junction. In the foreground, a PzKpfw III and SdKfz 251 advance.

tiple bases and moved as solid blocks of men) it is, in fact, rare to see more than a couple of Companies in action.

What many wargamers do is 'scale down' their organisation, so that in effect a Section of ten men *represents* a Platoon. On this basis, one box of Airfix figures could represent a 'Company' of four 'Platoons'. Half a dozen boxes of figures could then be formed up to represent a scaled-down 'Battalion' and the heavy support weapons allocated accordingly.

Although the precise organisation differed from army to army, and indeed from year to year during the course of the war, a fairly typical Battalion can be organised for any nationality for wargames purposes as follows:

First, three 'Rifle Companies' each of three 'Platoons' of ten men organised as described above for a Section. This is a total of 90 figures — nine NCOs, nine light machine-guns with two crew apiece, and 63 infantrymen with rifles.

Second, a 'Machine-Gun Company' of three 'Platoons', each 'Platoon' comprising three heavy machine-guns with one gunner, one loader and one ammunition carrier each, plus an NCO. Total 'Company' strength three NCOs, nine gunners, loaders and carriers — 30 figures.

Third, a 'Mortar Company' organised with 30 figures along exactly the same lines, but equipped with nine infantry mortars instead of machine-guns.

Fourth, an 'Anti-Tank Company' in which two of the 'Platoons' are equipped with three anti-tank rifles each, the third platoon having two light anti-tank guns (3.7 cm PAKs in the German army up to 1941, 5 cm weapons later; 2 pdrs in the British Army up to 1942, 6 pdrs later; 45 mms in the Russian Army early on, 57 mms later; 37 mms in the American Army, 57 mms later; 47 mms in the Italian and Japanese armies).

Fifth, a 'Headquarters Section' with the Battalion commander and his ADC, clerk, orderly, radio operator and a couple of 'runners' (messengers).

This organisation is, as you will appreciate, considerably simplified and does not specifically reflect the precise organisation of any particular army. If you want your troops organised exactly as in real life, you will need to refer to one or more of the many specialised books on your chosen army.

The above organisation can be used to represent an ordinary infantry Battalion or, by mounting each 'Platoon' in trucks or half-tracks, a Motorised (or Panzer grenadier) Battalion formed.

Armour

The armoured equivalent of the infantry Section — ie the smallest semi-independent fighting unit — was the Troop, of three to five tanks depending on nationality and year of the war.

Three Troops plus a command tank formed the average Company, and three Companies a Tank Battalion — 30 tanks in all. Once again, this is probably a greater number of tanks than you will ever require on one side of a wargames table using 1:76 scale models, so the same 'scaling down' process can reasonably be employed, giving you three 'Companies' of three tanks each plus a command vehicle to a wargames tank 'Battalion'. There were two to three Battalions to an average tank Regiment, one or two Regiments to a Division.

The type and organisation of tanks is important too. Taking the German Army first again, the standard organisation up to 1942 would give you two 'Companies' equipped with PzKpfw IIIs and one 'Company' equipped with PzKpfw IVs per Battalion. After this date the PzKpfw III began being withdrawn from front line service as the PzKpfw IV became available in larger numbers and the PzKpfw V 'Panther' introduced (1943). By 1944 the balance had swung so that the same 'Battalion' should have two

France 1940. British carrier and Light Tank Mk IV confer with a French Char B.

'Companies' equipped with PzKpfw IVs and one with PzKpfw Vs.

Tigers, and later the Tiger IIs or 'King Tigers', were normally formed in independent heavy tank Battalions.

SS formations normally received the new vehicles before the Wehrmacht — a situation which did not help the rivalry, distrust and, indeed, hatred which existed between them.

The British divided their armour into two, and later three, classes: Infantry tanks (Matilda, Valentine and Churchill); Cruiser tanks (Crusader and Cromwell); and Medium tanks (Grant and Sherman).

Within these groups the organisation was similar, a Regiment comprising three squadrons of tanks plus a headquarters element of four vehicles and a reconnaissance Troop. In the early days of the war two squadrons would be designated 'heavy' and equipped with Matildas (Infantry Tank Brigades), or A13s, Covenanters and Crusaders (Cruiser Tank Regiments). The third squadron in each case was designated 'light' and equipped with the tiny Light Tank Mk VI or, later, in the desert with the American 'Honey'. When the Grant, and later the Sherman, entered service, these equipped the 'heavy' squadrons of the Medium Tank Regiments. Similarly, the Matilda was gradually replaced by the Churchill after 1942 in the Infantry Tank Brigades, while the Cromwell replaced the Crusader in 1944.

The Americans, who standardised their equipment more than any other nationality except the Russians, only produced four main tank designs during the war, and organised them in 'light' and 'heavy' divisions equipped respectively with Stuarts, later Chaffees; and Lees, later Shermans.

The Russians organised their tanks into 'light', 'medium' and 'heavy' Battalions, Regiments and Brigades, the mediums taking the brunt of the fighting.

Beginning the war with the T-26 light tank, they introduced several new designs during the course of the war, including the T-50, T-40, T60, T-70 and T-80. These were primarily used as reconnaissance vehicles, and some were amphibious.

The initial Russian medium tanks were the BT-5 and later BT-7 which, although fast, were outclassed by German armour. They were rapidly replaced by the T-34 which, armed initially with a 76.2 mm gun and, from 1944 with an 85 mm weapon, was their main battle tank.

For infantry support the Russians used the KV-1 heavy tank throughout the war, although it was superseded after 1944 by the JS-II.

The Italians used 'medium' tank regiments equipped with Carro-Armato M11/39s and slightly superior M13/40s, plus 'light' tank regiments equipped with the tiny CV 3/35 tankettes.

The Japanese did not have tank divisions or any equivalent, doling out their Chi-Has individually as needed to perform specific tasks.

Readers who require more information on the tanks and other armoured fighting vehicles of World War 2 can do no better than to beg, borrow or buy copies of the three superb books published by Arms & Armour Press, *German Tanks of World War II,* by F. M. von Senger und Etterlin; *British and American Tanks of World War II,* by Peter Chamberlain and Chris Ellis; and *Russian Tanks 1900-1970,* by John Milsom. For information on the Italian army, a very useful book is Ralph Riccio's *Italian Tanks and Fighting Vehicles of World War 2* (Pique Publications). A further set of useful books is Liliane and Fred Funcken's four-volume *Arms and Uniforms: The Second World War.*

Artillery

There are so many different types

Trouble for the Allies. A troop of 5.5-inch guns ambushed by PzKpfw IVs and Panthers in France 1944. The guns are frantically trying to deploy for action while Australian infantry gallantly advance. On the right a Cromwell prepares to go down fighting.

and classes of artillery piece that it would be impossible to try to cover them all here. Primarily, there was Field Artillery, sub-divided into light, medium and heavy, which provided fire support for infantry attacks or was used to eliminate enemy troop or vehicle concentrations and plaster strongpoints; Anti-Tank Guns; Anti-Aircraft Guns; Mountain, and Airborne Guns. The latter two classes were light weight artillery pieces which could be dismantled easily either for carrying over rough terrain or for parachute drops. Field Guns can also be further sub-divided into Guns, which fired a shell on a fairly flat trajectory, and Howitzers, which used a high, plunging trajectory.

Artillery organisation, in general, followed the same pattern as infantry organisation. The smallest element was the Section, usually of two or three guns. Two or three Sections formed a Troop, two or three Troops a Battery and two or three Batteries a Regiment. The Battery is the equivalent of a Company and was called such in some cases, particularly in the care of anti-tank formations.

Artillery allocation varied widely. There were supporting Batteries included as integral parts of infantry and armoured divisions in most armies, but there were also independent Regiments which were frequently brigaded together into even larger formations to lay down concentrated barrages on particular points, or to precede attacks. How many artillery pieces you use in a wargame is therefore largely a matter of choice and the type of battle being fought. A typical infantry or armoured division would contain an integral Battalion at least, more usually a Regiment. However, since you will normally only be using one infantry Battalion in a wargame, and this is scaled down, a proportionate allocation would give one-ninth of an artillery Regiment's actual equipment. Say four Field Guns, 105 mms in the German and American armies, 25 pdrs in the British, 76.2 mms in the Russian, and 75 mms in the Italian and Japanese armies. Heavier weapons can be supplied independently as required

Of more importance in most wargames is the allocation of anti-tank guns. Simplifying again, a proportionate allocation would give you four weapons of the appropriate type (plus, of course, the integral anti-tank 'Company' already allocated to your infantry Battalion). These weapons would normally be heavier than those included in the Battalion anti-tank Companies, ie 5 cms, and later 7.5 cm PAK 40s and 8.8 cm PAK 43s in the German Army, 6 pdrs and later 17 pdrs in the British, 76.2 mms in the Russian, 57 mms in the American, 47 mms in the Japanese and Italian.

Anti-aircraft weapons were divided into two classes: light, quick-firing types of between 20 and 40 mm calibre; and heavy, long-range weapons such as the dreaded German '88'. In most forces these were allocated roughly on a two or even three to one basis, the light outnumbering the heavy. Thus in the German army you would not normally expect to have more than a couple of '88's on the wargames table at once, and for realism these would need to be accompanied by half a dozen 20 and 37 mm Flak guns mounted on trucks or half-tracks.

Recce units

Reconnaissance is a vital part of war and practically all formations of whatever nationality during World War 2 incorporated integral reconnaissance units. In the case of armoured formations these would take the form of troops or squadrons of light tanks and armoured cars; motorised infantry units would have half-tracks, motor cycles and sometimes armoured cars

Panzers advance. A German column comprising a Nashorn (foreground), SdKfz 7, PzKpfw IVF1, PzKpfw IVF2 and SdKfz 251 moves down a Russian road during 1943. A pair of Tiger 1s are visible on the hill at left, an SdKfz 232 recce vehicle on that at right.

Japanese Chi-Ha tank in a jungle wargames setting.

as well; while infantry units might have a few trucks or half-tracks but would more probably be mounted on bicycles!

Other units

Here the list is practically endless, and it is possible to build up a wargames army which includes signals, medical, engineer, catering and supply formations. Of these, those most likely to be required on a wargames table are signals units, to co-ordinate movement, transmit orders and receive battle reports; and engineer units, to construct or demolish field obstacles, bridge rivers, repair damaged vehicles and, when necessary, act as fighting infantry. Other formations are only really necessary in a wargames campaign.

Selecting an army

Once you have chosen your favourite nationality and decided to build up a wargames force, you will soon come to appreciate the need for some system of evaluating the comparative merits of opposing vehicle and weapon types. For example, suppose you have a German army of 1944 equipped with Panthers, Tigers, PzKpfw IVs, Puma armoured cars and PAK 40 anti-tank guns, while your opponent has an Allied army equipped with Shermans, Churchills, Daimler armoured cars and 17 pdr anti-tank guns. Neither the Sherman nor the Churchill were the equal, in either firepower or armour plate, of the Panther or Tiger, so if you allocate tanks on a 'one for you, one for me' basis, the Allied force will almost certainly be knocked out in pretty short order.

What is required is a points system for determining the relative value of different vehicles and weapons, so that, for example, a Tiger could be worth 120 points, while a Sherman is only worth, say, 70, so that the Allied player could have three Shermans to every two of your Tigers, thus redressing the balance.

The points values in the following list are based on a comparative assessment of firepower, armour plate and mobility (tanks and SP guns); armour piercing capability (anti-tank guns); load carrying ability (trucks and half-tracks — extra points have been added for armoured vehicles); and range and weight of shell (artillery).

Armour in the desert. German PzKpfw IVs versus British Matildas and Crusaders. (Note that in many of these photos the distances between vehicles have been compressed to minimise lighting and depth of focus problems. Wargames normally take place at greater ranges than these.)

German equipment

1 Tanks

PzKpfw I	10
PzKpfw II	15
PzKpfw 38(t)	20
PzKpfw III 37 mm	30
PzKpfw III 50 mm L/42	40
PzKpfw III 50 mm L/60	45
PzKpfw IVF1	60
PzKpfw IVF2	70
PzKpfw IVH	80
PzKpfw V Panther	100
PzKpfw VI Tiger 1	120
PzKpfw VI Tiger II	150

2 SP Guns

PzKpfw I Pak 47(t)	15
Marder II	50
Marder III	50
StuG III	65
Wespe	80
Hetzer	80
Jagdpanzer IV	90
Hummel	90
Nashorn	90
Jagdpanther	120
Elefant	150
Jagdtiger	200

3 Armoured cars

SdKfz 222	15
SdKfz 231/6-wheel	20
SdKfz 231/8-wheel	25
SdKfz 234/1	30
SdKfz 234/2 Puma	50
SdKfz 234/3	40
SdKfz 234/4	50

4 Transport

Kubelwagen	2
Field car	5
SdKfz 10	5
SdKfz 11	7
SdKfz 6	9
SdKfz 7	11
SdKfz 8	13
SdKfz 9	15
SdKfz 250	9
SdKfz 251	11
Maultier	9
Opel Blitz	5
Heavy truck	10

5 Anti-tank guns

Pak 35/36 37 mm	15
Pak 38 L/42 50 mm	25
Pak 38 L/60 50 mm	30
Pak 40 75 mm	50
Pak 41 75 mm	55
Pak 36(r) 76.2 mm	50
Pak 43 L/71 88 mm	90
sPzB 41 28 mm	20

6 Anti-aircraft guns

20 mm Flak 38	10
37 mm Flak 36/37	15
88 mm Flak 41	70

7 Artillery

75 mm	50
105 mm	70
150 mm	100

British equipment

1 Tanks

Light Mk VI	10
Matilda I	30
Matilda II	40
A10	30
A13	30
Crusader II	35
Crusader III	40
Cromwell	80
Valentine	40
Churchill III	70
Churchill VII	90
Tetrarch	25
Honey	30
Grant	60
Sherman	70
Sherman Firefly	90

2 SP Guns

Bishop	60
Archer	80
Deacon	50
Priest	75

3 Armoured cars

Daimler	20
Humber II	20
AEC II	30
Marmon Herrington	15

4 Transport

Jeep	2
Carrier	5
15 cwt truck	5
3-ton truck	10
Quad	10
DUKW	15

5 Anti-tank guns

2 pdr	10
6 pdr	40
17 pdr	60

6 Anti-aircraft guns

40 mm Bofors	20
3.7-inch	70

7 Artillery

25 pdr	60
5.5-inch	100

American equipment

1 Tanks

Stuart	30
Lee	60
Sherman	70
Chaffee	50
M26 Pershing	100

2 Tank destroyers

M10	80
M36	100
M18	70

3 Armoured cars

M8	25

4 Transport vehicles

M3 half-track	11
Jeep	2
15 cwt truck	5
3-ton truck	10
LRDG Chevrolet	7
Buffalo	20

5 Anti-tank guns

37 mm M3A1	20
57 mm M1	40

6 Anti-aircraft guns

M51 quad .50 MG	20
37 mm M1A2	15
90 mm M1	70

7 Artillery

75 mm (few only)	50
105 mm M2A1	70

Russian equipment

1 Tanks

T-26	15
T-38	10

T-50	20
T-70	30
BT-5	30
BT-7	40
T-28	40
T-34 76 mm	90
T-34 85 mm	110
KV-I	100
JS-I	110
JS-II	130
JS-III	160
T-35	80

2 SP Guns

SU-76	50
SU-85	80
SU-100	100
KV-II	110
JSU-122	120
JSU-152	150

3 Armoured cars

BA-10	30
BA-32	35
BA-64	40

4 Transport

GAZ 67	2
ZIS-5	5
GAZ-MM	10
GAZ-60 half-track	11

5 Anti-tank guns

45 mm	25
57 mm	35
76.2 mm	50
100 mm	90

6 Anti-aircraft guns

37 mm	15
76.2 mm	50
85 mm	65

7 Artillery

76.2 mm	50
85 mm	60
107 mm	70
122 mm	90
152 mm	110

Japanese equipment

1 Tanks

Chi-Ha	40
Te-Ke	20
Ha-Go	30

2 SP guns

Ho-Ni	50

The Japanese had no armoured cars and used little motorised transport, troops travelling on foot and guns being towed by horses or mules.

3 Anti-tank guns

47 mm Type 1	25

4 Anti-aircraft guns

75 mm Type 88	50
88 mm Type 99	80
120 mm Type 10	100

5 Artillery

75 mm Type 38	40
75 mm Type 90	50
105 mm Type 91	70

Italian equipment

1 Tanks

CV 3/35	10
M11/39	20
M13/40	35

2 SP guns

Semovente	35

3 Armoured cars

Autoblinda 40	15
Autoblinda 41	25

4 Transport

Fiat TL 37	5
Lancia 3 RO	10

5 Anti-tank guns

M35 47/32	25

6 Anti-aircraft guns

20 mm Breda	10
37 mm Breda	15
75 mm Ansaldo	50
90 mm Ansaldo	65

7 Artillery

75 mm (various)	40
105 mm Ansaldo	60

Taking this points system and working out a typical couple of examples, you can see that a German wargames tank battalion of 1944, with two 'Companies' of PzKpfw IVHs (six tanks), one 'Company' of Panthers (three tanks) and a command tank (another Panther, say), is worth a total of 880 points, while a comparable British mixed force comprising a squadron (three tanks) of Churchill IIIs and two squadrons of Shermans, plus command vehicle, is only worth 700 points. To make a balanced game, the British player should be allowed to choose additional tanks or guns totalling a further 180 points — he could have, for example, a couple of Fireflies.

Note that in the above list, all tanks and other AFVs are assumed to include a full crew complement; all trucks etc have a driver; and all guns have a normal detachment.

However, vehicles aren't all, and a similar points system is required for selecting balanced infantry forces. To begin with, therefore, we say that an ordinary infantryman is worth one point, an NCO two, field officer four and staff or general officer ten points.

Added to this you need to include the weapons carried, at the following rates: pistol or similar sidearm — ½; rifle or carbine — 1; sub-machine-gun

— 3; light machine-gun — 5; medium/heavy machine-gun — 10; light mortar — 10; heavy mortar — 15; anti-tank rifle — 10; bazooka, panzerfaust or similar — 15; grenades — ½ each; anti-tank grenade — 1 each.

Taking a typical infantry section as an example again, you would have one NCO with pistol and sub-machine-gun at 5½ pts; one private with light machine-gun at 6 pts; one loader with rifle at 2 pts; and seven privates with rifles and, say, two grenades each, at 3 pts — total 34½ points.

Additional equipment can also be allocated points values, such as a portable radio — 10; anti-personnel mine — 1; anti-tank mine — 3, and so on. If there is anything else you require, you can easily work out a reasonable points value for it yourself.

Right! Armed with this points system and some idea of what nationality you fancy, you are now in a position to build up a balanced wargames force for any period or theatre of the last war. Having carefully, we hope, painted up all your figures and vehicles, how do you fight with them? Read on . . .

five

Orders

The most important factor in wargames, yet the one most often forgotten or disregarded, is that of communication. Watch most wargames in progress, and it will become apparent that most troops are moved as if they were in communication with all their fellows by some form of telepathy. This, of course, is totally unrealistic, and to any wargamer who still plays thus I have one thing to say: no matter how realistic your rules for movement, firing, melee or morale, unless your communications system is equally well worked out, your 'games' are no more than just that. Playability is one thing. Selecting those rules which give you 'realism' while discarding even more important factors because they force you to commit moves to paper and forego a certain freedom of movement and choice, is *not* 'playing the game'.

So. To the point.

You, as wargamer in charge of the forces at your disposal, are acting out the role of a General or Field Marshal in effect. You have so many battalions, so many tanks and guns, under your control. On a wargames table you can see and manoeuvre them all as you will — if you will. But in a real life context, how would you, as senior commanding officer, actually communicate with the units under your command?

In World War 2 you have three basic permutations: word of mouth; field telephone; or radio. The first involves despatching an actual physical message by courier to a particular unit, with its attendant risks that the courier may be killed by a stray shell or, even worse, captured by the enemy so that your message not only does not reach its objective, but reveals to the enemy what you have in mind. The second involves soldiers actually being sent out, perhaps under bombardment, to lay telephone wires so that you can

A British Grant prepares to defend a fleeing convoy comprising Dorchester armoured command vehicle (ACV), Matador and 5.5-inch gun from a pair of German PzKpfw IVs and 5 cm anti-tank gun.

communicate with your subordinates. But the same stray shell which could kill your messenger could also cut the vital wire along which you are talking. The third involves communication — which could, and frequently was, be jammed by the enemy — on a particular radio frequency (called a 'net') with those units which possess receivers capable of accepting your call; receivers which, given the technology of the time (no transistors) were remarkably prone to breakdown; and how do you find a spare valve in the middle of a battle?

Communications are thus really vital, and any wargamer who neglects this aspect of warfare in his games cannot pretend that his battles bear any resemblance to reality.

However, it need not be as difficult as it sounds. Let us take the three basic means of communication in order and see how simple rules can be devised for their effect.

First, though, any wargamer planning a battle is in the position of a commanding officer and able to brief his officers beforehand. Thus, before a battle, general orders outlining the overall situation, giving the presumed enemy positions, and dictating basic tactical objectives, can be written out. These can be presumed to be in the hands of the officers commanding your brigades or battalions at the start of the battle, and it is fair to assume that, unless orders are changed or a unit is wiped out, they will be followed.

If the enemy dispositions are not as expected, however; or if you suffer an unexpected reverse at one point; or win a victory at another; you may wish to alter the basic tactics during the course of the battle.

Couriers

The first and most basic way in which you can issue alternative orders is by means of a physical messenger; a courier represented on the table by a figure (on foot, bicycle or motor cycle) who will be made to move according to the appropriate rate (see next chapter) and actually 'hand over' orders to the unit commander in the field.

This is the simplest of the three methods of transmitting orders and, in a wargame, can simply be done by writing out the new order, then moving your messenger until he reaches the unit concerned, when it will act upon the order.

Telephone

For orders to be transmitted by telephone, not only must the overall commander have a telephone or switchboard on the table, but the receiving unit must also have access to one. These are normally set up prior to the battle. Once set up, there are no problems except the chance of a stray shell cutting a wire, and for this reason the locations of telephone wires must be clearly indicated on a pre-battle map so that the possible effects of a shell burst can be assessed.

Radio

Radio communication relies on two sets — a transmitter and a receiver — operating on the same radio wavelength. For radio communication to be possible, a unit in the field must possess some form of portable radio, and this must be tuned in correctly.

At the start of a battle, all units can be assumed to be correctly tuned in, but, given the vagaries of valve-driven radios, the possibility of enemy jamming and, equally, the possibility of the enemy discovering your wavelength and transmitting false orders on it, anything can happen!

Reducing these three methods of communication to wargames level, we can say that messengers will deliver messages safely unless they come under enemy fire, in which case they may be killed (see chapter six). Similarly, telephone messages will arrive safely unless the wire or one end of the system has been eliminated by enemy action.

With radio, however, a more complicated system is needed. Initially, all radios are assumed to be in contact, and tuned to the correct frequency. But the enemy may have a radio system of their own, and may try to jam yours (or vice versa). This must be declared by the player concerned, and a dice thrown to see whether the jamming is successful. This was not an easy task, so if we say that the 'jamming' player needs to throw a 5 or 6 on his dice to jam successfully, it would seem about right. This must be done every Move the jamming is in progress. While jammed, no radio set can transmit or receive messages of any sort.

Similarly, the enemy (or you) may try to transmit false orders by radio. There is an even smaller chance of this succeeding, so if trying throw one dice again, and throw a 6 to succeed. Units receiving and accepting a false message *must* act on it (subject to initiative, see chapter seven) regardless of their previous orders.

To recapitulate: before moving any troops on the wargames table, each unit must have a set of basic orders and objectives written down on the player's order sheet (any old piece of paper will do); these orders can only be changed by means of the player transmitting a message via courier, telephone or radio to the unit concerned; if that order is not received, the unit will continue acting in accordance with its original orders until wiped out or altered by means of its own officer taking an initiative test with a successful result.

Time lag

It obviously takes time to write out an order, to read an order, and to transmit it to the troops concerned. In a wargames context let us say that it takes the CO one Move to write any order, whatever method it is to be transmitted by; one Move for the commander in the field to read and digest the order, by whichever means he receives it; and one further Move to transmit it to the men under his command.

Telephone and radio messages take one Move to transmit; couriers take the length of time it actually requires for them to move from point A to point B across the table, given the movement rates in the next chapter.

six

Movement

The next problem to be settled is that of movement. We have already established in chapter two that 1 millimetre on your wargames table represents 1 metre of actual ground, and that one battle Move will represent 30 seconds of real life activity. Using these as a yardstick, it is a simple matter to calculate, from historical records giving the appropriate march rates and vehicle performance figures, exactly how far each and every wargames figure or vehicle can travel per Move.

However, such performance figures are usually obtained under 'ideal' conditions — a flat, hard surface, good weather, and no obstacles. In a battle context, our troops have to contend with hedges, ditches, streams, woods, mud, rain, snow, ruined buildings and a variety of other obstacles to progress, not least of which is the fire being laid down by the enemy! The average rate of movement is thus going to be much slower than the ideal for most of the time.

What I have done in the following lists, therefore, is allocate a basic movement rate for troops and vehicles travelling across reasonably clear countryside; a second, faster, rate for troops and vehicles travelling along good metalled roads or other firm, flat surfaces (such as some areas in the desert); and a variety of movement penalties for troops traversing various obstacles. All measurements are in mm.

Infantry	Normal	Road	Broken ground and rubble	Woods	Jungle	Cross obstacle
Crawl	10	10	5	5	3	−½
Walk	30	40	20	20	10	−½
Run	50	60	30	30	10	−½

Matildas and a RHA 25 pdr attempt to fight off an assault by three PzKpfw IVs, North Africa 1941.

Broken ground and rubble means anything other than a reasonably flat surface. Woods are clear woods without much underbrush. Jungle is any dense, thickly overgrown wood or forest. It takes half a Move to cross an obstacle such as an ordinary hedge, fence, ditch, stream or low wall.

Other penalties

Movement by night is at half normal speed.

To enter or leave a vehicle or building takes a full Move.

To lie down or stand up takes a quarter of a Move.

Infantry under enemy fire must Run or Crawl.

Infantry carrying heavy machine-guns or mortars, or helping wounded comrades, move at half the above rates.

Movement may be in any direction, and all troops may be moved at once (ie, unlike Chess, you don't move one piece then wait for your opponent to move one — you both move together).

You can thus see that various combinations are possible within your 30 second wargame Move. Private Bloggs, for instance, could run 25 yards, which occupies half his Move; lie down, which takes him a quarter; and crawl for the remaining quarter of the Move, travelling a further 2½ yards.

While it is assumed that small arms will normally be loaded and ready to fire most of the time, heavy weapons such as machine-guns and mortars take time to set up or dismantle — in fact, a complete Move.

The actual mechanics of firing is dealt with in the next chapter, but it is important to realise at this stage that there is a limit on the physical activity permitted to each man during each wargame Move. You can't have a chap running and firing a heavy machine-gun from his hip, for example (except in war comics!). Each action is normally independent and takes up a finite portion of the total possible Move.

However, it would take so long and so slow down a wargame to take every tiny little factor into account — unless you were fighting a small scale skirmish with only half a dozen men — that most can reasonably be ignored. Thus there is no movement penalty for firing smallarms while moving (apart from heavy machine-guns and mortars, of course). What suffers, as you will see in the next chapter, is accuracy of aim.

Because tanks and other armoured fighting vehicles each have distinctive and important differences in their movement rates, they are treated individually in the following table.

Trucks and other soft-skin vehicles had comparable performances so no differentiation is made between types.

Starting up any vehicle's engine prior to moving takes a quarter of a Move.

Once again, all measurements are in mm.

Vehicle	**Normal**	**Road**	**Broken ground**	**Cross obstacle**
German				
PzKpfw I	130	336	90	−⅔
PzKpfw II	130	336	90	−½
PzKpfw III	150	336	100	−⅓
PzKpfw IV	140	336	95	−⅓
PzKpfw 38(t)	126	350	85	−½
Panther	205	380	150	−¼
Tiger I	168	310	115	−¼
Tiger II	150	310	110	−¼
PzKpfw I Pak 47(t)	130	336	90	−½
Marder II	130	336	90	−½
Marder III	130	336	90	−½
StuG III	200	336	140	−⅓

Field repairs on the wargames table. A Bergepanther recovery/engineer vehicle prepares to tow a broken-down Marder III (Gerald Scarborough).

Vehicle	Normal	Road	Broken ground	Cross obstacle
Wespe	168	336	110	−½
Hetzer	120	336	85	−½
Jagdpanzer IV	130	336	90	−⅓
Hummel	200	336	140	−⅓
Nashorn	200	336	140	−⅓
Jagdpanther	200	380	145	−¼
Elefant	90	168	70	−¼
Jagdtiger	135	310	95	−¼
SdKfz 222 A/C	336	672	170	−⅔
SdKfz 231 A/C	250	504	150	−½
SdKfz 234 A/C	360	714	200	−⅓
British				
Light Tank Mk VI	250	470	170	−⅔
Matilda I	80	110	60	−½
Matilda II	130	200	100	−⅓
A10	135	210	105	−⅓
A13	220	400	150	−⅓
Crusader II & III	200	336	140	−⅓
Cromwell IV	250	425	180	−⅓
Valentine	130	200	100	−⅓
Churchill	130	200	100	−¼
Tetrarch	300	540	180	−½
Honey	260	462	180	−½

'Aerial' view of part of a wargame set in the Ardennes during the Battle of the Bulge constructed by Airfix Magazine *reader S. Bloomfield. Visible are carriers, Shermans, Churchill, M3 half-track and a Priest, with a knocked-out Tiger in the left foreground.*

Vehicle	**Normal**	**Road**	**Broken ground**	**Cross obstacle**
Grant	200	336	140	−⅓
Sherman	200	336	140	−⅓
Sherman Firefly	200	336	140	−⅓
Bishop	130	200	100	−⅓
Archer	160	270	110	−⅓
Deacon	120	252	80	−½
Priest	200	336	140	−⅓
Daimler A/C	336	672	170	−⅔
Humber A/C	100	200	50	−⅔
AEC II A/C	200	390	120	−½
Marmon Herrington A/C	80	160	50	−½
DUKW	130	470	70	−½
American				
Stuart	260	462	180	−½
Lee	200	336	140	−⅓
Sherman	200	336	140	−⅓
Chaffee	270	470	170	−⅓
M26 Pershing	250	390	170	−¼
M10	200	336	140	−⅓
M36	250	336	150	−⅓
M18 Hellcat	330	590	200	−⅓
M8 A/C	360	720	180	−½
Buffalo	100	168	60	−⅓

Vehicle	Normal	Road	Broken ground	Cross obstacle
Russian				
T-26	200	294	135	−½
T-38	200	340	135	−½
T-50	200	425	135	−½
T-70	190	425	130	−½
BT-5	510	900	250	−⅓
BT-7	430	600	200	−⅓
T-28	160	180	130	−¼
T-34	336	400	190	−¼
KV-I	100	294	75	−¼
JS-I	160	310	110	−¼
JS-II	160	310	110	−¼
JS-III	160	310	110	−¼
T-35	160	250	120	−¼
SU-76	185	336	100	−½
SU-85	210	470	140	−¼
SU-100	210	470	140	−¼
KV-II	90	210	70	−¼
JSU-122	168	300	120	−¼
JSU-152	168	300	120	−¼
BA-10 A/C	210	470	110	−⅓
BA-32 A/C	210	470	110	−⅓
BA-64 A/C	330	672	200	−½
Japanese				
Chi-Ha	180	310	120	−⅓
Te-Ke	200	336	110	−½
Ha-Go	200	336	110	−½
Ho-Ni	220	378	150	−⅓
Italian				
CV 3/35	126	340	70	−½
M11/39	115	270	80	−⅓
M13/40	115	260	80	−⅓
Semovente	210	340	140	−⅓
AB 40 A/C	252	630	140	−½
AB 41 A/C	252	630	140	−½
All nationalities				
Light truck	360	670	230	−¾
Medium truck	320	580	210	−½
Heavy truck	230	500	160	−⅓
Light half-track	220	500	170	−½
Medium half-track	200	430	160	−⅓
Heavy half-track	170	370	140	−¼
Bicycle	70	100	40	−¼
Motor cycle	360	670	230	−¾

Amphibious vehicle movement through water 30 mm per Move

In all the above figures, it is important to realise that these are the *maximum* movement rates permitted during a wargame Move. There is nothing to prevent a man or vehicle moving less than this maximum.

Vehicles with fully revolving turrets may traverse their turrets while on the move and thus fire at targets to either side or rear, whereas those with fixed armament may only fire straight ahead.

Recce section. A pair of Kubelwagens and three BMW motor cycle combinations (Gerald Scarborough).

It would be possible, but time consuming, to enter precise figures for the length of time it takes each individual AFV to traverse its turret. For the purpose of these rules, I have taken as an average that it takes each tank one full Move to traverse through 360 degrees, thus a quarter of a Move to traverse through 90 degrees, and so on. No gun may be fired while being traversed. Those of you who wish to incorporate precise traverse times into your rules, however, can do so quite easily by referring to the various books in print on tanks and working out the times from these.

Infantry and vehicles may combine various types of movement within one wargame Move. For example, a Tiger tank may start its engine, which takes a quarter of its Move; travel a further quarter Move along a road — 77 mm; plough through a stone wall (a quarter of a Move); and travel its final quarter of a Move cross-country — 42 mm. It may simultaneously fire at any visible target on which it can bring its turret to bear.

Other movement

Apart from the elementary logistics of moving from A to B, there are several other battlefield tasks which can reasonably be classified under the heading of 'movement', such as digging a trench or man-handling an artillery piece. Some of these are listed here, though you will probably think of others yourself, and wish to incorporate them in the playing rules.

To dig a personal one-man foxhole takes an infantryman ten Moves (five minutes). To lay a mine (personnel or anti-tank) takes two Moves. To dig up a mine takes four Moves. To cut a one-man wide path through a section of barbed wire takes one Move. To man-handle a gun of up to 50 mm calibre, infantry move at half their basic Run rate; up to 80 mm calibre at half their Walk rate; and above this at half their Crawl rate.

Certain other special forms of movement are covered in the chapter on weapon firing, but the above are the most important ones which are likely to be needed in a wargame.

Right. At this stage of the book you know the essential ground and time scales; you know how to organise a basic wargames army; and you know how far and how fast your troops and vehicles may be moved on the table top. Having got them there, how do you make them fight?

seven

Combat

Combat, in miniature as in real life, can be resolved into three phases: firing, melee, and reaction. Under 'firing' we consider the effects not only of infantrymen firing their rifles and machine-guns at each other, but also of tank and anti-tank fire. artillery fire and anything else concerned with the projection of missiles from A to B. Under 'melee' we consider the simpler, but still vital, questions of infantry hand-to-hand combat, with bayonet, knife or fists. And under 'reaction' we consider the far more complex problems of troop behaviour under varying conditions on the battlefield.

Firing

This is principally a problem of logistics. All the data regarding weapon ranges, armour piercing ability, shell burst radii and other factors is readily available and easily assimilable into a tabular form reflecting the percentage probabilities of hits and casualties.

In a wargames context, what we are concerned with is quite simply, having moved man or vehicle X into such-and-such a position, how much damage can he or it reasonably be expected to inflict upon visible enemy targets? And how is this calculated and represented in miniature?

Firing can be sub-divided into several separate categories for easy reference: infantry versus infantry; infantry versus vehicles: AFVs versus AFVs; AFVs versus guns; guns versus AFVs; AFVs versus infantry; guns versus infantry. For our purposes, all vehicles, guns, tanks, etc, count as 'vehicles' in group two.

Infantry versus infantry

Before you can shoot at a target, you have got to be able to see it. This sounds obvious, but is frequently forgotten. An infantryman is a very small target even when he is standing in the open, so how much more difficult to see and hit is the man lying down, partially concealed behind a wall, or dug in in a foxhole? Similarly, although it is easier to spot a moving target, it is correspondingly more difficult to hit.

The following chart shows the distances at which a single figure on the wargames table may be spotted and correctly identified as one of the enemy. At maximum range it is necessary to throw a 5 or 6 on your dice to spot him; at medium range a 4, 5 or 6; and at short range a 3, 4, 5 or 6.

The crew of a PzKpfw IV surrender to a Crusader in the desert (Gerald Scarborough).

Figure is	Maximum range	Medium range	Short range
Standing in open	600 to 900	300 to 600	Below 300
Moving in open	700 to 1,000	400 to 700	Below 400
Crawling in open	400 to 600	200 to 400	Below 200
Standing in wood, underbrush, etc	200 to 300	100 to 200	Below 100
Moving in wood, underbrush, etc	250 to 350	150 to 250	Below 150
Crawling in wood, underbrush, etc	75 to 100	40 to 75	Below 40
Behind wall, etc, only head showing	100 to 150	50 to 100	Below 50
Lying down	125 to 200	75 to 125	Below 75
In foxhole	100 to 150	50 to 100	Below 50

All measurements are in metres (mm in a wargame). If the man being spotted fires his weapon add plus one to your dice throw, increasing your chances of seeing him. Alternatively, if he is partially hidden by smoke, deduct one, decreasing your chances of seeing him. Officers with binoculars add plus one to their dice score. (For details of smokescreens, see page 58.)

Once you have spotted a target, any or all of your own troops within range may fire at it. Weapon ranges are given in the following chart. I have not differentiated between individual types of infantry weapons as used by the various nations, since their capabilities were all roughly similar, although if you wish there is nothing to stop you working out a chart of ranges for each and every infantry weapon of the war. Finally, note that, although many weapons (heavy machine-guns, snipers' rifles, etc) had ranges greater than 1,000 metres, since you cannot spot a target at less than this, that is the maximum permissible range.

Weapon firing	Maximum range	Medium range	Short range
Throw grenade	20 to 30	10 to 20	Below 10
Pistol	35 to 50	20 to 35	Below 20
Sub-machine-gun	80 to 120	40 to 80	Below 40
Light machine-gun	400 to 600	200 to 400	Below 200
Medium machine-gun	600 to 900	300 to 600	Below 300
Heavy machine-gun	700 to 1,000	400 to 700	Below 400
Rifle/carbine	600 to 900	300 to 600	Below 300

When firing, throw a dice again to see whether you have hit the target exactly as before for spotting it: ie, a 5 or 6 at maximum range, 4, 5 or 6 at medium range, and 3, 4, 5 or 6 at short range.

Rifles and pistols are assumed to be fired four times during the 30 second Move at the target. If something — movement, for example — interferes with this, then obviously the chances of a hit are decreased further. So for every quarter of a Move during which firing does *not* take place, deduct one from your dice throw.

Automatic weapons, on the other hand, can throw out a continuous spray of bullets and may be traversed during the Move to fire at more than one target. The maximum traverse is 45 degrees either side of a straight line to the target — you can measure this with a protractor. When firing with a machine-gun at a group of men, therefore, you can throw dice for each possible target within the gun's cone of fire.

Finally, if the target figure(s) are crawling or running, the chances of hitting them are decreased, so deduct minus one from your dice throw. Similarly, if the firing figure is itself moving, its chances of achieving a hit are reduced still further, so deduct an-

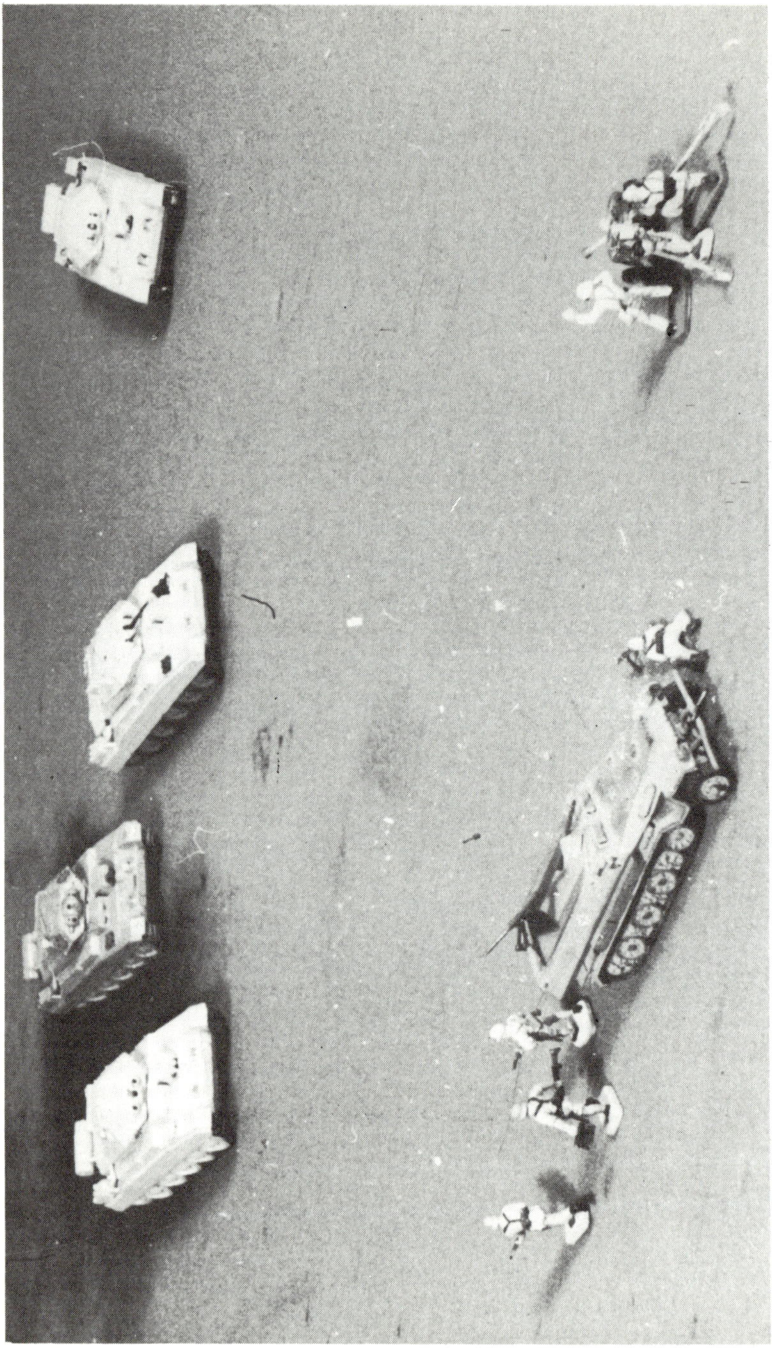

Part of a D-Day wargame photographed by Belgian Airfix Magazine *reader Doyle James. An Airfix gun emplacement is in the centre, flanked by other scratch-built strongpoints. In the foreground a Sherman and Matilda 'dozer advance up from the 'beach'.*

other one from the dice. And if the target is lying down or behind a wall or in a foxhole, deduct another two.

Having hit a target, the final step is to decide whether he is killed outright; badly wounded; slightly wounded; or whether in fact your bullet(s) only hit his back pack, helmet or something and he is unharmed.

Throw one final dice. If you get a 5 or 6 he is killed; 2, 3 or 4 he is wounded; 1 he is unharmed. If he is wounded throw the dice again: 4, 5 or 6 means he is badly wounded; 1, 2 or 3 means he is slightly wounded.

Men who are badly wounded may not move or fire for eight Moves, and thereafter crawl at half normal rate. They will only fire their weapons in self defence, with a minus two off their dice throw.

Men who are slightly wounded may not move or fire for four Moves. They may then crawl or walk (but not run) at half normal rate, and may fire their weapons with minus one being deducted from their dice throws.

Let's now run through a typical sequence to get everything straight. Get a couple of figures out of a box of Airfix soldiers and put them down on a table (or the floor) 700 mm apart (measuring the distance with a steel tape measure or similar). Let's call the figures A and B (or Hans and Jock, if you prefer). Hans throws a dice to see whether he spots Jock — go on, do it. Say you get a 5 or 6 — this means he has spotted him. Jock, on the other hand, might only get a 2, which means that at this range he has not spotted Hans.

Hans can now fire at Jock but not vice versa. Let us say he is carrying an ordinary rifle. He thus needs to get another dice throw of 5 or 6 to hit Jock. Go on — throw that dice. Say it comes up less than 5 — he has

A German 5 cm anti-tank gun prepares for action against a surprise attack by Crusader IIs.

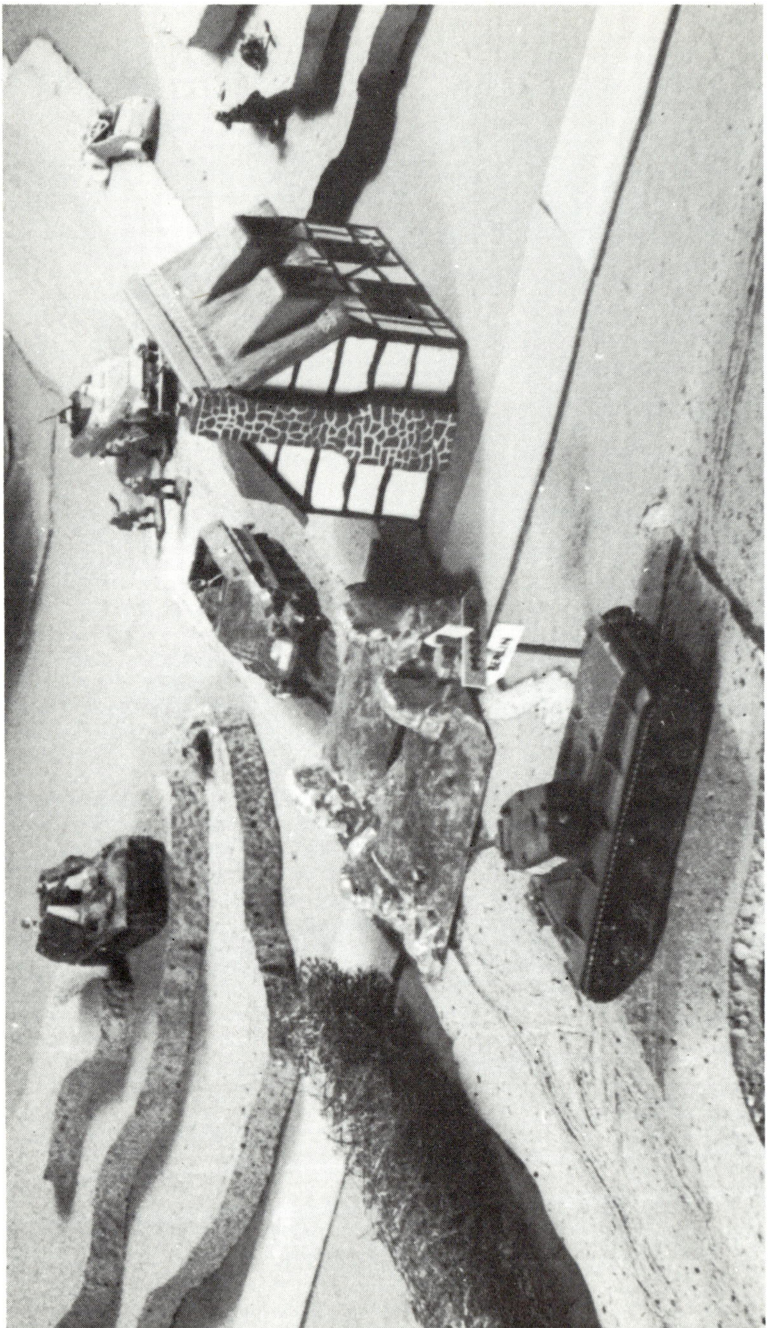

The first Tigers arrived in North Africa after Alamein and saw action in Tunisia. Here they are observing a British position while a mine detector team clears the ground behind the barbed wire.

missed. But Jock now knows there is someone firing at him.

You can now move the two figures. Let us say that Hans runs to try to close the range, while Jock sensibly falls flat on his face to present a smaller target while he tries to spot his enemy. Lying down takes Jock a quarter of his Move, during which Hans runs 12 yards closer to him. Because Hans is moving, this time Jock only needs a 4, 5 or 6 to spot him, plus one because he has been fired at, so in fact a 3, 4, 5 or 6 would do.

Hans fires again, but this time he is minus two on the dice because he is running while firing, and minus a further two because Jock is lying down. At 650 yards range he does not, therefore, stand any chance of hitting Jock at all.

Assume Jock has spotted Hans this Move. He cannot hit him either because at this range he needs a 5 or 6 for a hit, minus one because Hans is moving, and minus another one because he spent a quarter of his Move lying down.

And so it goes on. You can follow this example through yourself until one or both figures are incapacitated. And by starting out using just two figures you will grasp the essentials of moving and firing far more quickly than if you plunged immediately into a full-scale game. Sprinkle a few obstacles around the table to give cover, get a friend or relative to take over control of one of the figures, and you have a first class miniature war-game going.

Before moving on to the next section, one weapon which requires special rules is the grenade, which explodes and sends splinters of steel flying in all directions for several yards. You must therefore throw dice for casualties on all figures within a ten yard radius of the target figure, as well as for the effect of the grenade on that figure itself.

Infantry versus vehicles

This section can be divided conve-

Trouble! A Russian KV-1, which has already knocked out the Kubelwagen visible in the right background, blocks a bridge. Panzer grenadiers from the two SdKfz 251s run for cover, while a Wespe SPG courageously advances to tackle the Soviet monster.

niently into two parts: infantry fire against unarmoured ('soft-skin') vehicles; and against armoured vehicles. The latter required special weapons since armoured vehicles were mainly impervious to small arms fire, and is tackled in the pages on anti-tank fire below.

Infantry fire against soft-skin vehicles is performed in much the same way as infantry versus infantry fire, except, of course, that a truck — be it a jeep or a tank transporter — is a much bigger target than a man, and thus easier to hit when it is stationary. It is also a faster moving, and therefore more difficult, target when on the move.

The following chart shows the ranges at which vehicles may be seen, and once spotted, at which they may effectively be fired upon. The dice-throwing procedure is exactly as for infantry fire — a 5 or 6 at maximum range, 4, 5 or 6 at medium and 3, 4, 5 or 6 at short range.

Nominally, jeeps, kubelwagens and similar small four-wheeled vehicles count as 'light' vehicles, six-wheeled types as 'medium' and eight or more wheeled types as 'heavy'.

If a stationary vehicle has been camouflaged with netting or similar, deduct two from the dice to lessen the chance of its being seen.

Having spotted a vehicle, throw for a

Observation

Type of vehicle	Maximum range	Medium range	Short range
Light truck, in open, stationary	700 to 1,000	400 to 700	Below 400
Light truck, in open, moving	900 to 1,200	600 to 900	Below 600
Medium truck, in open, stationary	900 to 1,200	600 to 900	Below 600
Medium truck, in open, moving	1,100 to 1,500	800 to 1,100	Below 800
Heavy truck, in open, stationary	1,100 to 1,500	800 to 1,100	Below 800
Heavy truck, in open, moving	1,300 to 1,700	900 to 1,300	Below 900

hit using the Weapon firing table on page 39, but add plus one to your dice throw at all ranges if the target is stationary, and deduct minus one at all ranges if it is moving. Vehicles may be observed but not fired upon at ranges over 1,000 metres. (This may sound pointless, but having once observed a vehicle, you can then keep a check on where it goes, even if your opponent tries to hide it, and consequently shoot at it the moment it does come within range. If you wish, until a vehicle is observed and correctly identified by your opponent, you could use a dummy piece — for example, a small block of wood — to represent it.)

Having hit a vehicle, throw another dice to assess the damage. If firing a rifle or other non-automatic weapon, it requires a 6 to kill the driver, whereupon the vehicle will crash and be written off, or a 5 to puncture a tyre, which forces the vehicle to halt for four Moves during which the crew change the wheel. With automatic weapons, a 5 or 6 will kill the driver, a 3 or 4 will puncture a tyre. Any other scores do not harm the vehicle.

If the vehicle is carrying troops, throw a dice for each occupant (up to a maximum of four with non-automatic weapons), 5 or 6 counting as a hit. Dice for effects of hits as in infantry versus infantry fire.

Grenades hitting soft-skin vehicles automatically immobilise them, and all occupants must test for injuries.

Tank and anti-tank fire

This is slightly more complicated because of the widely differing thicknesses of armour plate on different AFVs, and because of the varying capabilities of anti-tank guns, but the basic principles are the same.

Before you can fire, however, you must first observe the tank, throwing a dice and referring to the Observation chart in the infantry versus vehicles section. Light tanks count as light trucks, medium as medium trucks and heavy as heavy trucks.

Now we have three factors to consider: first, whether or not a hit is achieved at all; second, what is the armour thickness at the point of impact; and third, whether the type of gun firing is capable of piercing that thickness at that range.

To score a hit, first of all, refer to the following chart and throw a dice. In each case the dice score given is the *minimum* required at each range. (ie the score given or anything higher is a hit; anything lower is a miss).

Left France 1944. A troop of Churchills advances past a ruined farmhouse.

Desert champions. A scratch-built PzKpfw III with long-barrelled 5 cm gun alongside a PzKpfw IVF2 (Gerald Scarborough).

Close-up of the PzKpfw III which was built up on the Airfix StuG III chassis (Gerald Scarborough).

Gun firing is	Range in mm 0-100	101-250	251-400	401-600	601-800	801-1,200	1,201-1,500	1,501-1,800
Below 20 mm calibre	2	3	4	5	6			
21-37 mm	Any	2	3	4	5	6		
38-50 mm	Any	Any	2	3	4	5	6	
51-77 mm	Any	Any	Any	2	3	4	5	6
Above 78 mm	Any	Any	Any	Any	2	3	4	5

Deduct minus one from the dice score if the target vehicle is moving; deduct minus two if it is 'hull down' behind a ridge, or dug-in.

Now, what precisely have you hit? Tanks and other AFVs normally had different thicknesses of armour plate on different parts of their anatomy, the thickest armour being on the turret and hull front, the thinnest at the rear.

The following chart lists most of the vehicles you will encounter on the wargames table, together with different values for their armour thickness at different points ranging from 'A' (very thick and strong) to 'F' (very thin). These values take into account not only the actual thickness of the armour, but the angle at which it was sloped.

Vehicle	Armour classification Turret Front	Turret Sides	Turret Rear	Hull Front	Hull Sides	Hull Rear
German						
PzKpfw I	F	F	F	F	F	F
PzKpfw II	E	F	F	E	F	F
PzKpfw 38(t)	E	E	E	E	F	F

Vehicle	Armour classification					
	Turret Front	Turret Sides	Turret Rear	Hull Front	Hull Sides	Hull Rear
PzKpfw IIID-J	E	E	E	E	E	E
PzKpfw IIIL-M	C	E	E	D	E	D
PzKpfw IVD-F1	D	E	E	E	F	F
PzKpfw IVF2	D	E	E	D	E	F
PzKpfw IVH	C	D	D	C	E	F
Panther	A	D	D	B	D	D
Tiger 1	B	C	C	B	C	C
Tiger II	A	B	B	A	B	B
PzKpfw I Pak 47	F	F	F	F	F	F
Marder II	F	F	F	E	F	F
Marder III	F	F	F	E	F	F
StuG III	D	E	E	D	E	E
Wespe	F	F	F	E	F	F
Hetzer	C	E	F	C	E	E
Jagdpanzer IV	B	D	F	D	E	F
Hummel	F	F	F	E	F	F
Nashorn	F	F	F	E	F	F
Jagdpanther	B	C	D	B	D	D
Elefant	A	C	C	A	C	C
Jagdtiger	A	B	C	A	B	C
SdKfz 222	F	F	F	F	F	F
SdKfz 231	F	F	F	F	F	F
SdKfz 234	—	—	—	E	F	F
Puma	E	F	F	E	F	F
British						
Light Mk VI	F	F	F	F	F	F
Matilda I	C	D	D	C	D	E
Matilda II	B	C	C	B	C	E
A10	E	F	F	F	F	F
A13	E	F	F	E	F	F
Crusader	D	F	F	D	F	F
Cromwell	B	C	C	B	C	D
Valentine	C	F	F	C	F	F
Churchill	B	C	C	B	C	D
Tetrarch	F	F	F	F	F	F
Honey	E	E	E	D	E	E
Grant	C	D	D	D	E	E
Sherman	C	D	E	D	E	E
Bishop	E	F	F	C	F	F
Archer	E	F	F	C	F	F
Deacon	F	F	F	—	—	—
Priest	F	F	F	D	E	E
Daimler A/C	F	F	F	F	F	F
Humber A/C	F	F	F	F	F	F
AEC Mk II A/C	E	F	F	E	F	F
Marmon-Herrington A/C	F	F	F	F	F	F
American						
Stuart	E	E	E	D	E	E
Lee	C	D	D	D	E	E
Sherman	C	D	E	D	E	E

Vehicle	Armour classification Turret Front	Turret Sides	Turret Rear	Hull Front	Hull Sides	Hull Rear
Chaffee	F	F	F	F	F	F
Pershing	B	D	D	B	E	E
M10	D	F	F	E	F	F
M36	F	F	F	E	F	F
Hellcat	F	F	F	F	F	F
M8 A/C	F	F	F	F	F	F
Buffalo	F	F	F	F	F	F
Russian						
T-26	E	F	F	E	F	F
T-38	F	F	F	F	F	F
T-50	D	E	E	D	E	E
T-70	C	F	F	D	F	E
BT-5	F	F	F	F	F	F
BT-7	F	F	F	E	F	F
T-28	C	F	F	E	F	F
T-34/76	B	D	D	D	D	D
T-34/85	A	B	B	C	D	D
KV-I	B	C	C	A	C	C
KV-II	B	C	C	A	C	C
JS-I	A	B	B	A	B	C
JS-II	A	A	B	A	B	C
JS-III	A	A	A	A	B	B
T-35	F	F	F	E	F	F
SU-76	E	F	F	E	F	F
SU-85	B	D	D	B	D	D
SU-100	B	D	D	B	D	D
JSU-122	A	B	B	A	B	C
JSU-152	A	B	B	A	B	C
BA-10 A/C	F	F	F	F	F	F
BA-32 A/C	F	F	F	F	F	F
BA-64 A/C	F	F	F	F	F	F
Japanese						
Chi-Ha	F	F	F	F	F	F
Te-Ke	F	F	F	F	F	F
Ha-Go	F	F	F	F	F	F
Ho-Ni	E	F	F	F	F	F
Italian						
CV 3/35	F	F	F	F	F	F
M11/39	F	F	F	F	F	F
M13/40	E	F	F	E	F	F
Semovente	F	F	F	E	F	F
AB 40 A/C	F	F	F	F	F	F
AB 41 A/C	F	F	F	F	F	F

One of the most revealing factors to emerge from a study of these lists is the armour thicknesses used on most German and Russian tanks compared with those of other nations — no other nationality has any 'A' class armour at all. However, to continue:

Having ascertained that you have secured a hit on a vehicle, extend your tape measure in a straight line from

Russia, 1943. Part of a wargame scene constructed by Argentinian Airfix Magazine *reader Jorge Wilburn. Visible are Tigers, StuG III assault guns, a scratch-build Nashorn, and '88. In the background a KV-2 burns.*

the firing gun to the target tank. This will show you whether your projectile hits front, side or rear. To discover whether you have hit the hull or turret, throw a dice, 5 or 6 giving you a turret hit, 1, 2, 3 or 4 a hull hit. Vehicles which are 'hull down' can only be hit on the turret, of course, so if you throw a 1 to 4 in this case your shell is assumed to have ploughed harmlessly into the ridge or embankment in front of the target.

The next stage is to find out whether the type of gun firing can, at that particular range, pierce the armour of the target at whichever point it has hit. The following table lists all the main categories of tank and anti-tank gun on the left, with the ranges along the top. The armour classifications given represent the maximum thickness which can be penetrated at the given range — ie 'D' means that that gun can pierce D, E or F class armour, but not A, B or C.

For those of you who aren't sure which type of gun is carried by any given tank or AFV, there is a breakdown after this table.

Gun calibre/ type				**Ranges in mm**				
	0-100	101-250	251-400	401-600	601-800	801-1,200	1,201-1,500	1,501-1,800
20 mm	D	E	F					
37 mm/2 pdr	D	E	F	F				
45/47 mm	D	D	E	E	F	F		
50 mm L/42	C	D	D	E	E	F		
50 mm L/60	C	C	D	D	E	E	F	
57 mm/6 pdr	C	C	D	D	E	E	F	

Gun calibre/ type	Ranges in mm							
	0-100	101-250	251-400	401-600	601-800	801-1,200	1,201-1,500	1,501 1,800
75 mm L/24	D	D	E	E	F	F		
75 mm L/43-48	A	B	B	C	C	D	E	F
75 mm L/70	A	A	B	B	C	C	D	E
76.2 mm/17 pdr	A	B	B	C	C	D	E	F
85 mm	A	A	B	B	C	C	D	D
88 mm L/56	A	A	B	B	C	C	D	D
88 mm L/71	A	A	A	B	B	B	C	C
90 mm	A	A	B	B	C	C	D	D
100 mm	A	A	A	B	B	B	C	C
122 mm	A	A	A	A	B	B	B	C
128 mm	A	A	A	A	A	B	B	B

20 mm — carried by *German* PzKpfw II, SdKfz 222, SdKfz 231, SdKfz 234/1; *Italian* AB 41.

37 mm — carried by *German* PzKpfw III, PzKpfw 38(t); *British* Matilda I and II, A10, A13, Crusader I and II, Valentine I-VII, Tetrarch, Grant (turret), Humber IV, Daimler, Marmon-Herrington Mk IV; *American* Stuart (Honey), Lee (turret), M8; *Russian* T-38, BA-10, BA-32; *Japanese* Te-Ke, Ha-Go; *Italian* M11/39.

45/47 mm — carried by *German* PzKpfw I Pak 47(t); *Russian* T-26, T-50, T-70, BT-5, T-35 (two turrets); *Italian* M13/40.

50 mm L/42 — carried by *German* PzKpfw III.

50 mm L/60 (Pak 40) — carried by *German* PzKpfw III, Puma.

57 mm/6 pdr — carried by *British* Crusader III, Churchill III, Valentine VIII-X, AEC Mk II, Deacon; *Japanese* Chi-Ha.

75 mm L/24 — carried by *German* PzKpfw IV, StuG III; approximate equivalent *British* Grant; *American* Lee; *Japanese* Ho-Ni; *Italian* Semovente.

75 mm L/43-48 — carried by *German* PzKpfw IVF2-H, Marder II and III, StuG III, Hetzer, SdKfz 234/4; approximate equivalent *British* Cromwell, Churchill, Sherman, Valentine XI, AEC Mk III; *American* Sherman, Chaffee.

75 mm L/70 — carried by *German* Panther, Jagdpanzer IV.

76.2 mm/17 pdr/3-inch — carried by *German* Marder II and III; *British* Archer, Sherman Firefly; *Russian* BT-7, T-28, T-35, T-34/76, KV-I, SU-76; *American* M10, Hellcat.

85 mm — carried by *Russian* T-34/85, SU-85.

88 mm L/56 — carried by *German* Tiger 1.

88 mm L/71 — carried by *German* Tiger II, Jagdpanther, Elefant, Nashorn.

90 mm — carried by *American* Pershing, M36.

100 mm — carried by *Russian* SU-100.

122 mm — carried by *Russian* JS-I, II and III, JSU-122.

128 mm — carried by *German* Jagdtiger.

Italy, late 1943. A convoy of M3 half-tracks led by a Grant tank has run into trouble from a pair of Panthers. Help is on the way in the form of a Sherman and mortar-carrying half-track on the right.

Apart from these main weapons, there were a variety of others carried by some vehicles. Several carried machine-guns only (German PzKpfw I, British Light Mk VI, Italian CV 3/35, AB 40, Russian BA-64, American Buffalo). The German Wespe carried a 105 mm gun/howitzer and the Hummel a 150 mm gun, while the British Bishop mounted a 25 pdr and the Priest a 155 mm gun. The Russian KV-II and JSU-152 both mounted 152 mm guns. See the following section on artillery for fire from these weapons.

Finally, assuming that you have observed and hit a tank with a gun which is capable of penetrating its armour plate at that particular range, you need to assess damage. If you want a quick, easy game you could simply say that whenever this happens, the vehicle is permanently knocked out, but if you want a more realistic assessment of results, use the following chart each time you achieve a successful hit.

Hit on turret — throw a dice: a score of 5 or 6 lifts the turret clean off, killing all occupants; a 3 or 4 penetrates the turret killing all crew in the tank apart from the driver (and hull machine-gunner, if present); the tank can therefore still move but cannot fight; a 1 or 2 damages the turret traverse mechanism, fixing the turret facing in the direction it was pointing when hit for the remainder of the game. The tank will therefore have to line itself up like as SPG (SPGs, of course, are not themselves affected by a score of 1 or 2).

Hit on hull — throw a dice: a score of 5 or 6 penetrates the tank, exploding the petrol and ammunition and killing all the crew; a 3 or 4 penetrates the hull — throw a dice for each crew member, requiring a 6 to save him; and a 1 or 2 breaks a track, which takes the crew six Moves to fix.

Tank guns do not, of course, all fire at the same rate. Generally speaking, the larger the gun, the slower it was capable of firing, and although individual rate of fire tables could be prepared for each type, it is simpler to classify them by weight. A notable exception was the German 8.8 cm dual-purpose gun, which could in fact fire faster than most other types regardless of size. In my wargames, I simply use the following figures: guns of up to 50 mm calibre, *and* the '88, can fire four times per Move, 51 to 76 mm, three times; 77 to 100 mm, twice; and over 100 mm, once.

Normally all fire will be against one target — ie if you don't knock it out first shot, you carry on throwing dice for hits within the same Move until, hopefully, you do. But if you succeed with your first shot, you may wish to re-align your gun on a different target.

Tank turret traverse times have already been covered; to elevate or depress an anti-tank gun on to a target closer or further away than the first target, takes half a Move; to traverse a gun on to a new target to the left or right of the first one takes half a Move for every ten degrees of arc.

Let's now follow a typical battle example through, taking what are probably the two most popular AFVs as adversaries — an Allied Sherman and a German Tiger 1. Set them up on opposite sides of a table, or on the floor, say 700 mm apart as you did for the infantry. The Sherman is a medium tank and the Tiger heavy, so as far as observation goes the Sherman requires a dice score of 3, 4, 5 or 6 to spot the Tiger (heavy tank at short range) while the Tiger needs a 4, 5 or 6 to spot the Sherman (medium tank at medium range).

At 700 yards the Sherman, with a 75 mm gun, needs a dice score of 3 to hit the Tiger, while the Tiger with its 88 mm gun only needs a 2, or better of course. Let's suppose they both hit each other frontally — remember, all movement and firing is supposed to be simultaneous. A dice score of 1 to 4 will now give you a hull hit in each case, 5 or 6 a turret hit. Let's say you get two hull hits, this being the most likely result (after all, it's a bigger target).

At 700 yards range, the Sherman's 75 mm gun can only penetrate 'C' class armour, but the Tiger's frontal armour is 'B', so the Sherman cannot

The Airfix Waterloo farmhouse, especially with the roofs removed as here, will fit well into any World War 2 battlefield. Additional irregular gaps can easily be cut into the brickwork, and the kit can also be cut up to form three separate buildings.

harm it. At the same range, however, the Tiger's 88 mm gun can also penetrate 'C' class armour, and since the Sherman's hull front is 'D' class, the Tiger's hit will be successful. Throw a final dice to assess damage . . . and remember in future not to send Shermans up against Tigers from the front! Had the Sherman in this case been approaching from the Tiger's flank, where its armour plate is only 'C' class, it could have been successful.

So to recap, the procedure in tank or anti-tank gun fire is: **1** — dice for observation; **2** — dice for a hit; **3** — assess whether the firing can can penetrate at the range; **4** — assess damage.

Going back to our example briefly, the Tiger, having eliminated its target first shot, could now continue to traverse its turret and fire at a second target within the same Move, if required. It would be quite possible for it to knock out another Sherman in the same Move, in fact, though this would take a fair bit of luck.

Firing infantry anti-tank weapons against tanks and other AFVs follows exactly the same procedure, the main difference being that infantry anti-tank weapons have far shorter ranges than pukka tank and anti-tank guns. Follow the stages as before, but use the following charts under stages (2) and (3):

Dice for hit

Weapon	**Range in mm** 0-25	26-50	51-100	101-200	201-300	301-400	401-500
Anti-tank rifles							
German (PzB 38/39)	2	3	5	6			
British (Boys)	2	3	4	5	6		
Russian (PTRD 1941)	2	3	4	4	5	5	6
Japanese (Type 97)	2	3	4	5	6		
'Bazookas' etc							
German (Panzerfaust)	2	4	5	6			
(Panzerschreck)	2	3	4	5	6		
British (PIAT)	2	4	5	6			
American (Bazooka)	2	3	4	5	6		

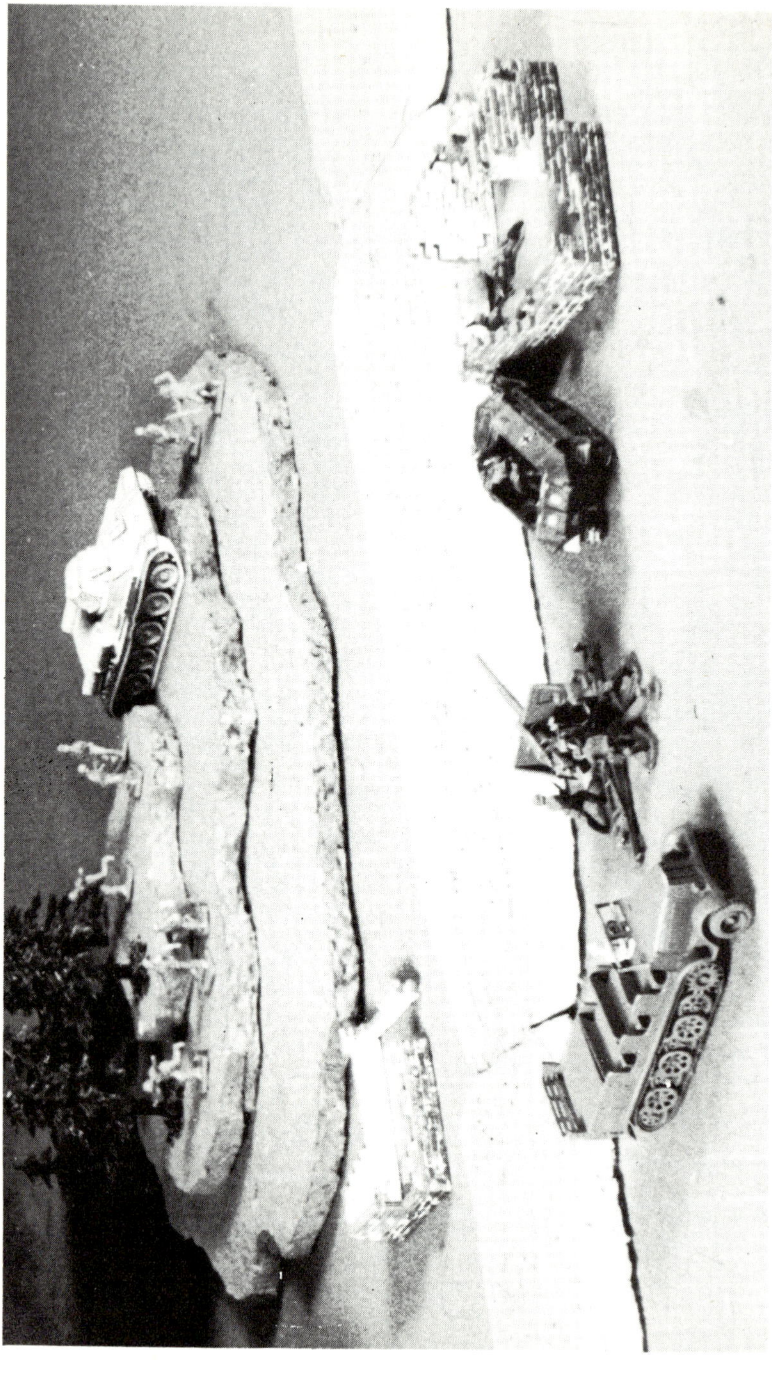

Armour penetration at range hit

Weapon	Range in mm						
	0-25	26-50	51-100	101-200	201-300	301-400	401-500
PzB 38/39	D	D	E	F			
Boys	D	E	E	F	F		
PTRD 1941	C	D	D	E	E	F	F
Type 97	D	E	E	F	F		
Panzerfaust	A	A	A	A			
Panzerschreck	A	A	A	A	A		
PIAT	A	A	A	A			
Bazooka	A	A	A	A	A		

In other words, if you can achieve a hit with a hollow-charge weapon (Panzerfaust, Bazooka, etc) a 'kill' is guaranteed within the limitations of the weapon's range. Vehicles with hull or turret anti-bazooka skirts (eg German PzKpfw IVH) suffer no effective damage from hits on those protective plates, the hollow-charge shell exploding harmlessly between the skirt and the main armour.

We have now ascertained our methods of firing infantry weapons at each other; infantry weapons against soft-skin and armoured vehicles; and armoured vehicles and anti-tank guns against armoured vehicles. The next problem is artillery fire — both high explosive (HE) shells from tanks against guns, plus all the other permutations: HE fire against infantry, against soft-skin vehicles, against AFVs, and against each other. We also have the further problems of direct (line of sight) and indirect fire. So let's make a start.

Artillery fire

We have already discussed the basic types of gun briefly under 'artillery' in Chapter three; what we have to do now is assess their effectiveness against different types of target. Anti-tank gun fire against AFVs has been covered, but that's about all. How does a tank fire back on an anti-tank gun? First, it has to be able to fire HE — something most British guns were unable to do during the early part of the war, and one of the reasons for the Germans' successes.

HE fire differs from AP (armour piercing) in one most important respect. With the latter, the principal consideration is normally muzzle velocity (ie the speed at which the shot is travelling when it leaves the gun barrel. This, together with the shell's weight, determines its impact velocity, or the speed at which it is travelling when it hits its target, and its effectiveness against armour plate. That is why, for example, a 75 mm shell from a long gun barrel (eg L/70) is normally more effective than one from a short barrel (eg L/24), even though they are both the same weight). With HE fire, on the other hand, what is important is the weight of explosive carried by a given shell. Unlike the armour piercing shot, this will explode with the same effectiveness regardless of range. This, in turn, explains why the hollow charge type of projectile is equally effective regardless of range, up to the maximum distance it can be fired: it depends on its weight of explosive, not speed, to pierce its target's armour.

What all this means in wargames terms is that firing HE is far simpler

Part of a wargame depicting action in Russia during the autumn of 1942. An unwary T-34/76 has come under fire from a hastily deployed German '88. Russian infantry, who were riding on the back of the T-34 (a standard Russian practice) are rapidly scurrying for shelter from the machine-gun fire laid down by the SdKfz 251.

than firing AP shot, even though there were a far larger number of types of field gun and howitzer in use during World War 2 than there were anti-tank guns. Field artillery fire can be subdivided into two categories: direct, and indirect. The former applies when a gun or battery is firing against a visible target, the latter when a gun is being directed on to a target which it cannot see by an observer. The effect of the shell fired depends, of course, on its size regardless of whether it is directly or indirectly fired, so in a wargames context we are primarily concerned with whether or not a hit is achieved. And equally obviously, it is easier for a gun crew to hit a target which they can actually see than one on to which they have been directed by an observer.

However, to get down to brass tacks, in a wargames context we can say that direct fire can be directed against any objective on the table which would in real life be visible to the gun firing; indirect fire can either be fire over a hill or other obstacle, or from *off the table.* Given the extremely ling ranges of some World War 2 artillery pieces — frequently several miles — this is a standard device used by many wargamers. In both cases, indirect fire relies on your having an observer with a radio actually on the wargames table and in a position to see the target. He radios instructions to the gun battery, observes the fall of shot, and issues corrections until a hit is achieved. The effectiveness of a given shell, of course, is the same whatever the range, and whether it is fired directly or indirectly.

The following table lists the main types of gun capable of firing HE, including those tank and anti-tank guns which had HE capability; the weight of shell fired; maximum effective range in metres (mm on the wargames table, of course); rate of fire in terms of rounds per wargame Move; and the shell burst radius (ie area within which damage and casualties would normally be caused). This list is not exhaustive, since there were literally hundreds of different types of gun in use, but it does cover those most commonly encountered in combat and most likely to be required on the wargames table.

All figures given in this table are 'rounded' to an even number so far as possible. Where two or more guns of a particular calibre existed in any army, characteristics have been taken for the most commonly used, which are

For the ambitious, amphibious wargames present a real challenge. This is a scene from Terry Wise's enactment of Operation Sealion — the projected German invasion of Britain in 1940 — which has been described in full in Airfix Magazine.

not necessarily the best.

Rates of fire are for guns firing continuously at one target throughout their Move. To correct a setting and raise or lower the aim takes half a Move. To traverse a field gun takes half a Move for every ten degrees of arc. To manhandle a field gun, refer to page 37.

Ranges are really immaterial, since any target on a wargames table is going to be within the reach of any field gun, but are included for those of you who wish to use 'off table' fire.

To see whether you achieve a hit with a field gun, throw a dice as follows, adding plus one in each case when firing direct. 6=direct hit on target. 5=hit within effective shell burst range of target (does not affect AFVs). 4=hit 50 metres behind target. 3=hit 50 metres in front of target. 2=hit 50 metres to left of target, and 1=hit 50 metres to right of target. Any troops or vehicles hit accidentally suffer just as though they were the real target.

If you miss first shot, throw a dice again after elevating, depressing or traversing the gun, adding plus one to your dice throw. Keep adding plus one each shot until a hit is achieved. When firing against a stationary target, all subsequent shots are automatically hits. When firing against a moving target, you have to dice each time as if it was a first shot, regardless of whether a hit has been achieved.

Weapon	Weight of shell (lb)	Maximum effective range	Rounds per Move	Shell burst radius
German				
37 mm	1.5	5,500	5	10
50 mm	4.5	2,750	5	15
75 mm	12	3,800	5	20
88 mm	21	4,100	4	35
105 mm	33	10,700	3	35
150 mm	96	15,500	2	60
British				
25 pdr	25	12,250	5	30
3.7-inch	20	5,500	4	40
4.5-inch	50	18,250	3	50
5.5-inch	100	14,600	2	70
7.2-inch	200	15,200	1	110
American				
75 mm	14.6	8,750	5	20
105 mm	33	11,150	3	35
155 mm	96	15,000	2	60
8-inch	200	17,000	1	80
Russian				
76.2 mm	14	11,150	5	25
85 mm	20	2,750	4	30
122 mm	48	11,800	3	45
152 mm	88	12,900	2	60
Japanese				
75 mm	12.5	11,000	5	20
105 mm	31.5	10,500	3	25
150 mm	68	19,200	2	35

Weapon	Weight of shell (lb)	Maximum effective range	Rounds per Move	Shell burst radius
Italian				
75 mm	14	10,250	5	25
105 mm	38	17,500	3	35
149 mm	101	17,500	2	60
All nationalities				
50 mm mortar	2	500	8	15
80 mm mortar	7.5	1,000	6	20
120 mm mortar	35	1,500	4	40

(All figures for mortars approximate, based on typical examples.)

What precisely do all these figures mean? In a wargame they mean that every man within the shell burst radius stands a good chance of being killed, and to determine casualties I fall back on the time-honoured method of 'saving throws'. In other words, throw one dice for each man within the shell burst radius, requiring a 6 to save him.

Direct shell bursts on soft-skin vehicles automatically destroy them, and you need to throw a 'saving' dice for each occupant. Unarmoured guns count as soft-skin vehicles against HE fire.

HE shells falling close to soft-skin vehicles (ie within burst radius, but not a direct hit) stand a 50 per cent chance of destroying them. Throw a dice with 4, 5 or 6 destroying the vehicle, 1, 2 or 3 saving it.

HE shell bursts on AFVs are ineffectual up to 105 mm calibre. Any shell of that size or larger hitting even the best-protected AFV stands a good chance of concussing the crew, breaking tracks and radio aerials, and generally causing havoc to the vehicle hit. When firing HE at AFVs therefore, with guns of over 105 mm calibre, throw a dice for effect. 1 and 2 has no effect, 3 means a track has been broken and will have to be repaired, 4 means that the AFV's radio aerials have been shot away and it cannot receive further orders, or transmit reports, while 5 or 6 means that the crew is concussed and cannot function for two Moves.

Smoke

All artillery pieces are capable of firing smoke shells, and most tanks also had smoke grenade dischargers on their turrets. Laying down a smokescreen is a useful way of concealing movement — the enemy may know *something* is happening, but it will take him longer to identify precisely what, because his visibility is cut in half.

Tanks equipped with dischargers may fire smoke at the same time as they are firing their main and/or secondary armament (don't forget, most tanks were equipped with one or more machine-guns as well as their main gun). Artillery pieces may opt to lay down smoke instead of firing HE.

Smoke from burning buildings and vehicles can also obscure visibility on the battlefield and needs to be taken into account.

Smoke from tank dischargers forms a screen extending for 20 metres in all directions around the vehicle and lasts for four Moves when there is no wind, three Moves in light wind, two Moves in medium wind and one Move in high wind. Smoke clouds move in the direction of the wind at the rates specified below.

Smoke from artillery shells last the same lengths of time, and cover the following areas: shell up to 80 mm calibre — 100 by 50 metres; shell above 80 mm calibre — 150 by 75 metres.

Smoke can be represented on the wargames table by pieces of cardboard of the appropriate length, cut in an uneven, rounded shape at the top, or by cotton wool.

Smoke from burning buildings or vehicles produces a cloud extending 25 yards around the building or vehicle, but tends to rise vertically and thus does not obscure visibility any further. Tanks and buildings set on fire will normally continue burning for at least an hour (120 Moves) which will usually be throughout a game.

Visibility within a smokescreen is half normal (eg if you are observing a vehicle 700 metres away which is surrounded by smoke, you can see it, but your 'effective visibility' is only 350 metres, so you cannot correctly identify it). A similar ruling could be devised to take into account fog or mist on the battlefield.

Wind

At the beginning of the wargame allocate each side and corner of your table a number, from *two* to nine (you'll se why in a moment). Now throw two dice and add their scores. The resulting number shows the section on the table from where the wind is blowing (10, 11 and 12 are null — throw again if your score totals any of these).

To determine wind strength, throw one more dice at the same time: 1=no wind, 2 and 3=light wind, 4 and 5=medium wind, and 6=strong wind. Light wind blows smoke 25 metres per Move; medium 50 metres; and strong 100 metres.

Right. We've now covered firing all the main types of weapons normally encountered on the wargames table. You could go on to add rules for firing heavy artillery (eg coastal and railway guns), rockets (eg Russian Katysuahas), and for ground attack by aircraft and opposing anti-aircraft fire. There isn't room to go into all of these here, I'm afraid, but a good way of resolving aerial attack and anti-aircraft fire is outlined in my earlier book, *Tank Battles in Miniature 2: A wargamers' guide to the Russian Campaign 1941-45,* also available from Patrick Stephens Ltd. So far as ground action is concerned, however, you can now fire any type of weapon at any target in sight, which does, in fact, cover the basics of World War 2 wargaming. However, you also need to be able to fight hand-to-hand battles, or melees, on occasion; and also judge troop reactions to what is going on around them — after all, even though they're plastic, they ought to have simulated human feelings: filled with elation when they winning, and quite the reverse when placed in an untenable position.

In World War 2 wargaming, neither of these considerations is as important as it is in other historical periods, partly because the majority of troops during 1939-45 were better trained and disciplined than previously, hence more likely to follow orders literally, in some cases, to the last man; and partly because melees were comparatively rare.

Melees

These can be tackled quite simply and satisfactorily when they do occur by a straight dice throw, the one scoring highest winning and killing his opponent. To make things a little more realistic, you also ought to add one to the dice score of a veteran or elite soldier — a guardsman, member of the Waffen SS, or similar — giving him increased chances of winning; and deduct one from the score of a raw, green conscript. It *would* be possible, but impractical, to draw up tables for the effectiveness of one type of hand-to-hand weapon against another (eg knife, bayonet, rifle with bayonet attached, rifle used as club, entrenching tool, bare fists, etc), and if you want to take your games this far then you are perfectly welcome to do so. A very good guide to this form of combat is provided in Donald Featherstone's book *Skirmish Wargaming,* also available from Patrick Stephens Ltd.

Reaction

This is the last combat factor which needs consideration, and in many ways is the most important. To give an extreme example, a well-fed soldier from an elite regiment, well armed, well rested, well led and securely entrenched in a strong position, with plenty of support around and behind him, is more likely to function well than a half-starved, tired, badly led and ill-equipped man in a hopeless position with no support.

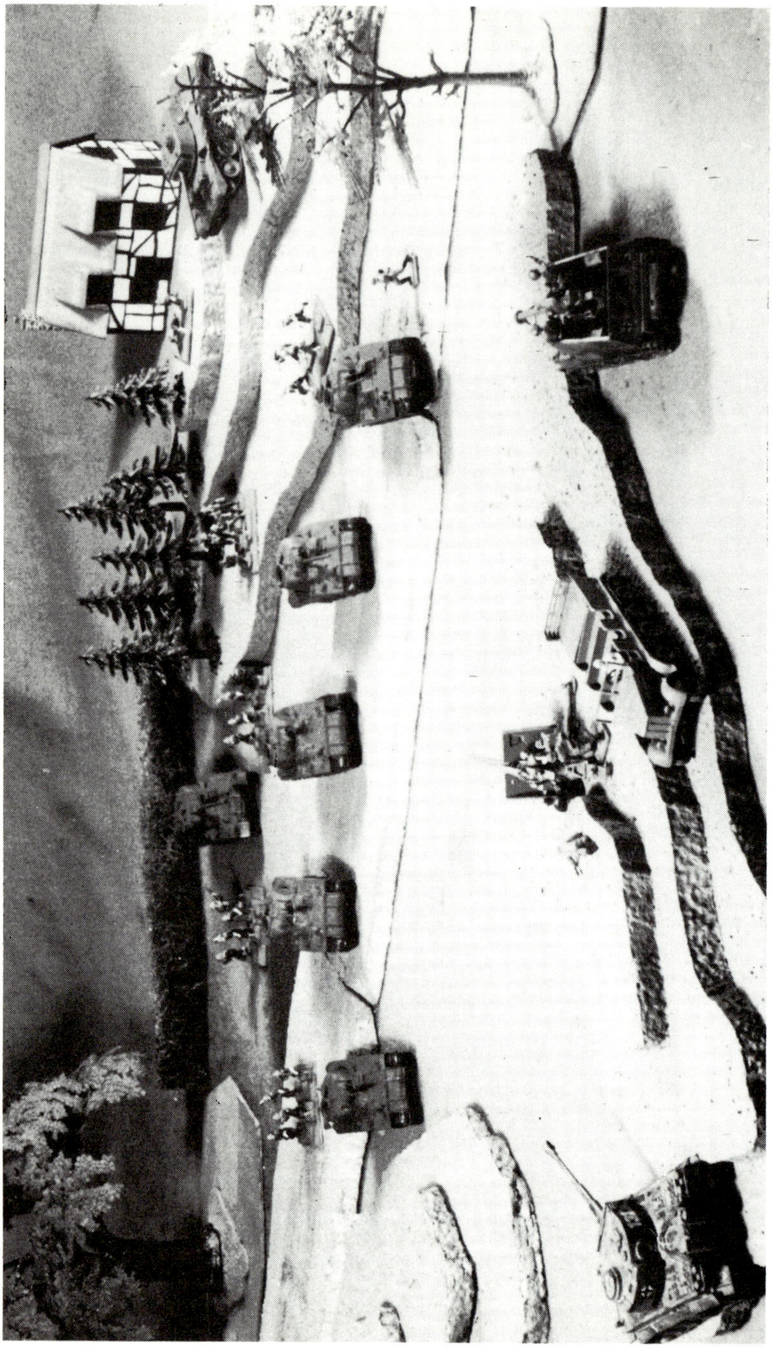

You can, if you wish, assume that every man in your army is an invincible hero, and dispense with reaction rules. This certainly simplifies matters and speeds up a wargame, but it is not truly realistic. If you want to assess reaction, then, use the following table. It can be applied either to individual model soldiers, or to whole groups which can be assumed to react as one, such as tank and gun crews, etc.

A reaction test should be taken whenever a man or unit comes under fire for the first time; whenever it suffers its first casualties; whenever an officer is killed; whenever a supporting man or unit is forced to run by a reaction test result; whenever charging or being charged (ie moving into physical contact with the enemy); and in any other circumstances which common sense tells you might worry the troops concerned — for example, a surprise attack from an unexpected direction.

To assess the morale of an individual or unit, first throw one dice, then add or deduct from your score according to the following table.

Many more factors could be taken into account, but these are the most essential. If you want to incorporate further ones for state of health, tiredness, etc, then I'm not going to stop you! But they *do* slow up a game.

Anyway, reaction results are as follows: a total of 4 or more means that the man/unit being tested will follow its orders and continue the task in hand. With any score below 4, throw another dice. A 1 or 2 means that the man/unit will run as quickly as possible in the most direct available line away from the nearest enemy unit until circumstances alter so that its morale is raised, or until it runs right off the table. A 3 or 4 means that the man/unit will retire, in good order, back on the closest supporting friendly unit, then hold firm. A 5 or 6 means it will halt where it is *or* retire at will, but will not advance or charge.

Troops dug in, hull down, inside buildings or protected by some other form of cover, automatically add +1 to their scores. Tank and AFV crews do *not* add anything for their vehicle's armour plate.

Conscripts, untried troops, etc		−1
Veterans, elite troops, guardsmen, etc		+1
Unit at full strength		+1
Unit at ¾ to full strength	Does not apply to	—
Unit at ½ to ¾ strength	individuals; instead, deduct −1	−1
Unit at ¼ to ½ strength	on lightly wounded men,	−3
Unit at below ¼ strength	−3 on badly wounded	−5
Officer killed this Move		−3
For each officer within unit killed or wounded so far		−1
For each supporting friendly man/unit*		+1
For each threatening enemy man/unit**		−1

* To count as support, an individual man must be within 25 mm of the figure being tested; or within a multiple of 25 mm equal to the number of men in the unit being tested (eg ten men in unit being tested — all friendly units of equal or greater strength within 250 mm count as support).

** To count as threatening, an individual man must be within 25 mm of the figure being tested; or within a multiple of 25 mm equal to the number of men in the unit being tested (eg ten men in unit being tested — all enemy units of equal or greater strength within 250 mm count as threatening).

Last ditch! Tiger, 8.8 cm gun and Wespe prepare to sell themselves dearly against attacking Allied Shermans. There is a Firefly in the background and a Cromwell on the right.

Initiative

Earlier I said that all orders would be obeyed without question, subject to initiative. What this means is that a good officer who receives an order which is patently silly in the light of his immediate situation may decide on his own responsibility to ignore it. However, occasions when this happens are rare in most modern armies because of the discipline, and normally an officer will obey his superior regardless of what he privately thinks of them! In fact, the better the officer, the more likely he will be to carry through orders. So how do we take this factor into account in a wargame?

First, either before or during the course of a wargame, each officer needs to be assigned a 'leadership factor', which reflects his initiative, courage, imagination, education, training and personality. Throw one dice for him, with the following results:

1 — below average officer with extremely cautious streak. Will always query every new order before acting on it, and even after receiving confirmation will delay two Moves before putting it into effect. 2 — below average officer with cautious streak. Will query any order which would put his unit into a more exposed position than that it currently occupies. Will delay one Move before complying with any new order.

3 — average officer. Will obey orders immediately under normal circumstances, but will query any order which wculd put his unit into a more exposed position before complying. 4 — average officer who will obey every new order immediately without question.

5 — above average officer who will normally obey every order instantly and without question, but will act at own discretion if the situation appears hazardous. 6 — above average officer who acts as (5) but also reports his plans to his superior officer when disobeying orders, so that at least the latter knows the position.

Above average officers will also query false orders received from the enemy over the radio before acting on them.

Any unit commanded by a below average officer deducts one from its morale score when taking a reaction test; a unit with an above average officer adds one to its morale score.

This takes care of most elements you will need to get a first-class World War 2 wargame going. Obviously, you could go on practically endlessly adding extra rules and refinements, but as I said at the beginning, to me the *game* is the important thing, and you will find that using these rules, although they are far simpler than many, you will normally get the same result at the end of a game as you would using more complicated ones. After all, it's the individual player's skill in positioning and manoeuvring his troops, guns and vehicles which wins battles.

You will find that you play better if you know the rules thoroughly. Nine times out of ten the average but experienced player will be able to beat a better player who is not so familiar with the rules, simply because his reactions are faster and he is not continuously thinking 'now can I do so-and-so, and if so, how?'. Wargaming is just like Chess in this respect at least. So my advice is to read through this book at least once more, from cover to cover, and practice each particular element of the playing rules with just a couple of figures and vehicles, before you set up a full-scale game. And may all your battles be victories.

Bibliography

Military equipment

WW2 Fact Files at £1.20 each paperback are an ideal source of information on the guns and other weapons of all nationalities. Published by Macdonald & Jane's, titles so far released of interest to wargamers are as follows: *Anti-tank weapons, Machine guns, Mortars and Rockets, Anti-aircraft guns, Infantry, Mountain and Airborne Guns, Light and Medium Field Artillery, Self-Propelled Anti-tank and Anti-aircraft Guns* and *Heavy Artillery.*

For information on the tanks and other vehicles, Chris Ellis & Peter Chamberlain's *British and American Tanks of World War 2,* John Milsom's *Russian Tanks 1900-1970* and F. M. von Senger und Etterlin's *German Tanks of World War 2,* all published by Arms & Armour Press, are beyond compare.

Profile publications booklets on a variety of tanks and other weapons are a useful source of additional information, if you can find them in the shops, as are the excellent series of Bellona booklets published by Argus Press. The latter include 1:76 scale four or five-view plans as well as technical data, making them invaluable for modellers.

The American WE Inc range, distributed in Britain by Patrick Stephens Ltd, includes some useful titles, the best of which are *Tank Data 2, Tank Data 3, German Combat Weapons of World War 2, Japanese Combat Weapons of World War 2,* and *German Tank and Anti-tank.*

For information on 'soft-skin' vehicles the best source is Bart Vanderveen's *Observers Fighting Vehicles Directory, World War 2.*

World War 2 campaigns

There are so many books on the different campaigns that to try to list even a representative selection would be doing injustice to the field. However, bearing cost in mind one of the finest series is the Purnell *History of the Second World War* series, inexpensive paperbacks on all the important battles and campaigns. These are usually stocked by model shops as well as W. H. Smiths and station bookstalls.

Wargaming

An excellent recent book for the World War 2 wargamer is John Sandars' *An Introduction to Wargaming* (Pelham Books), while for those of you with more specific interests, Donald Featherstone's *Tank Battles in Miniature 1: A wargamers' guide to the Western Desert Campaign 1940-1942* and, dare I say, my own *Tank Battles in Miniature 2: A wargamers' guide to the Russian Campaign 1941-1945* are both very useful.

Index

The following is not a complete index to every fact in this book. It is simply a guide to the pages on which particular playing rules appear, to facilitate reference during the actual course of a wargame. To make things even easier, many players photocopy the appropriate rules, leaving out the descriptive text, on to a few sheets of paper, and you may wish to do the same. Do remember, however, that these copies must be for personal use only, and *may not be sold* to other wargamers.